# Estimator's Man-Hour Manual

## on . . . Heating
## Air Conditioning
## Ventilating
## Plumbing

## Man-Hour Manuals by John S. Page

Estimator's Equipment
Installation Man-Hour Manual/2nd Edition

Estimator's General Construction
Man-Hour Manual/2nd Edition

Estimator's Man-Hour Manual on Heating,
Air Conditioning, Ventilating, and Plumbing/2nd Edition

Estimator's Piping Man-Hour Manual
(co-author)/3rd Edition

Estimator's Electrical Man-Hour Manual (co-author)

## Other Books by John S. Page

Cost Estimating Manual for
Pipelines and Marine Structures

Estimator's Manual of Equipment and Installation Costs

# Estimator's Man-Hour Manual on Heating, Air Conditioning Ventilating and Plumbing

**Second Edition**

John S. Page

Gulf Publishing Company
Book Division
Houston, London, Paris, Tokyo

**Estimator's Man-Hour Manual on Heating, Air Conditioning, Ventilating, and Plumbing**



Library of Congress Catalog Card No. 61-10176
ISBN 87201-364-2

Second Edition
First Printing, March 1978
Second Printing, November 1982

To my wife, Celesta,

# PREFACE

This book is not intended to hold anything new for the well-seasoned mechanical estimator whose ability, know-how and knowledge in this field is the product of years of schooling, actual competitive bidding, hard knocks and time-consuming analyzation of both good and bad estimates. Its main intention is to assist the partially experienced plumbing and mechanical estimator by affording a basis for arriving at reasonable dollar value for direct labor operations.

The following direct labor manhour tables are the product of hundreds of time and methods reports and preplanned studies coupled with actual cost of various operations both in the field and fabricating shops. These came from many varied types of plumbing and mechanical installations for large commercial and industrial projects ranging in cost from $10,000 to $100 million.

After careful analysis of these many reports we found that a productivity of seventy (70) per cent was a fair average for all crafts that might be involved in a normal plumbing and mechanical contract. The direct labor manhours throughout this manual are based on this percentage.

You will find no cost as to materials, equipment usage, warehousing and storage, fabricating shop set-up or overhead. If a material take-off is available, this cost can be obtained at current prices from vendors who are to furnish materials. Warehousing and storage, fabricating shop set-up, equipment usage and overhead can readily be obtained by a good estimator who can visualize and consider the job schedule, size and location. These are items which can and must be considered for the individual project.

Before an attempt is made to apply the following direct labor manhour tables we caution the estimator to be thoroughly familiar with the introduction on the following pages entitled "Production and Composite Rate," which is the key to this type of estimating.

# INTRODUCTION

## Production and Composite Rate

Herein lies the correct method for the application of the many manhour tables that follow.

There should be sound reasoning and understanding to back up a monetary unit before it is applied to an item for labor value. The best reasoning that we know is manhours based on what we call productivity efficiency coupled with production elements.

After comparison of many projects, constructed under varied conditions, we have found that production elements can be grouped into six different classifications and that production percentages can be classified into five different categories.

The six different classifications of production elements are:

1. GENERAL ECONOMY
2. PROJECT SUPERVISION
3. LABOR RELATIONS
4. JOB CONDITIONS
5. EQUIPMENT
6. WEATHER

The five ranges of productivity efficiency percentages are:

| Type | Percentage Range |
|---|---|
| 1. Very low | 10 through 40 |
| 2. Low | 41 through 60 |
| 3. Average | 61 through 80 |
| 4. Very good | 81 through 90 |
| 5. Excellent | 91 through 100 |

Since there is such a wide range between the productivity percentages, let us attempt to evaluate each of the six elements, giving an example with each, and see just how a true productivity percentage can be obtained.

1. GENERAL ECONOMY: This is nothing more than the state of the nation or area in which your project is to be constructed. The things that should be reviewed and evaluated under this category are:

   a. Business trends and outlooks
   b. Construction volume
   c. The employment situation

   Let us assume that after giving due consideration to these items you find them to be very good or excellent. This sounds good, but actually it means that your productivity range will be very low. This is due to the fact that with business being excellent the top supervision and craftsmen will be mostly employed and all that you will have to draw from will be inexperienced personnel. Because of this, in all probability, it will tend to create bad relationship between Owner representatives, Contract supervision, and the various craftsmen, thus making very unfavorable job conditions. On the other hand, after giving consideration to this element you may find the general economy to be of a fairly good average. Should this be the case, you should find that productivity efficiency tends to rise. This is due to the fact that under normal conditions there are enough good supervisors and craftsmen to go around, they are satisfied, thus creating good job conditions and understanding for all concerned. We have found, in the past, that general economy of the nation or area where your project is to be constructed, sets off a chain reaction to the other five elements. We, therefore, suggest that very careful consideration be given this item.

2 As an example, to show how a final productivity efficiency percentage can be arrived at, let us say that we are estimating a project in a given area and after careful consideration of this element, we find it to be of a high average. Since it is of a high average, but by no means excellent, we estimate our productivity percentage at seventy-five (75) percent.

PROJECT SUPERVISION: What is the calibre of your supervision? Are they well-seasoned and experienced? What can you afford to pay them? What supply do you have to draw from? Things that should be looked at and evaluated under this element are:

a. Experience
b. Supply
c. Pay

Like general economy this too must be carefully analyzed. If business is normal, in all probability, you will be able to obtain good supervision, but if business is excellent the chances are that you will have a poor lot to draw from. Should the contractor try to cut overhead by the use of cheap supervision he will usually wind up doing a very poor job. This usually results in a dissatisfied client, a loss of profit, and a loss of future work. This, like the attachment of the fee for a project, is a problem over which the estimator has no control. It must be left to Management. All the estimator can do is to evaluate and estimate his project accordingly.

To follow through with our example, after careful analysis of the three (3) items listed under this element, let us say that we have found our supervision will be normal for the project involved and we arrive at an estimated productivity rate of seventy (70) percent.

3. LABOR CONDITIONS: Does your organization possess a good labor relations man? Are there experienced first class satisfied craftsmen in the area where your project is to be located? Like project supervision, things that should be analyzed under this element are:

   a. Experience
   b. Supply
   c. Pay

   A check in the general area where your project is to be located should be made to determine if the proper experienced craftsmen are available locally, or will you have to rely on travelers to fill your needs. Can and will your organization pay the prevailing wage rates?

   For our example, let us say that for our project we have found our labor relations to be fair but feel that they could be a little better and that we will have to rely partially on travelers. Since this is the case, we arrive at an efficiency rating of sixty-five (65) percent for this element.

4. JOB CONDITIONS: What is the scope of your project and just what work is involved in the job? Will the schedule be tight and hard to meet, or will you have ample time to complete the project? What kind of shape or condition is the site in? Is it low and mucky and hard to drain, or is it high and dry and easy to drain? Will you be working around a plant already in production? Will there be tie-ins, making it necessary to shut down various systems of the plant? What will be the relationship between production personnel and construction personnel? Will most of your operations be manual, or mechanized? What kind of material procurement will you have? There are many items that

could be considered here, dependent on the project; however, we feel that the most important items that should be analyzed under this element are as follows:

a. Scope of work
b. Site conditions
c. Material procurement
d. Manual and mechanized operations

By a site visitation and discussion with Owner representatives, coupled with careful study and analysis of the plans and specifications, you should be able to correctly estimate a productivity percentage for this item.

For our example, let us say that the project we are estimating is a completely new plant and that we have ample time to complete the project but the site location is low and muddy. Therefore, after evaluation we estimate a productivity rating of only sixty (60) percent.

5. EQUIPMENT: Do you have ample equipment to complete your project? Just what kind of shape is it in and will you have good maintenance and repair help? The main items to study under this element are:

   a. Usability
   b. Condition
   c. Maintenance and repair

   This should be the simplest of all elements to analyze. Every estimator should know what type and kind of equipment his company has, as well as what kind of mechanical shape it is in. If equipment is to be obtained on a rental basis then the estimator should know the agency he intends to use as to whether they will furnish good equipment and good maintenance.

   Let us assume, for our example, that our company equipment is in very good shape, that we have an ample supply to draw from and that we have average mechanics. Since this is the case we estimate a productivity percentage of seventy (70).

6. WEATHER: Check the past weather conditions for the area in which your project is to be located. During the months that your company will be constructing, what are the weather predictions based on these past reports? Will there be much rain or snow? Will it be hot and mucky or cold and damp? The main items to check and analyze here are as follows:

a. Past weather reports
b. Rain or snow
c. Hot or cold

This is one of the worst of all elements to be considered. At best all you have is a guess. However, by giving due consideration to the items as outlined under this element, your guess will at least be based on past occurrences.

For our example, let us assume that the weather is about half good and half bad during the period that our project is to be constructed. We must then assume a productivity range of fifty (50) percent for this element.

We have now considered and analyzed all six elements and in the examples for each individual element have arrived at a productivity efficiency percentage. Let us now group these percentages together and arrive at a total percentage:

| Item | Productivity Percentage |
|---|---|
| 1. General economy | 75 |
| 2. Production supervision | 70 |
| 3. Labor relations | 65 |
| 4. Job Conditions | 60 |
| 5. Equipment | 70 |
| 6. Weather | 50 |
| Total | 390 |

Since there are six elements involved, we must now divide the total percentage by the number of elements to arrive at an average percentage of productivity.

$390 \div 6 = 65$ percent average productivity efficiency

At this point we caution the estimator. This example has been included as a guide to show one method that may be used to arrive at a productivity percentage. The preceding elements can and must be considered for each individual project. By so doing, coupled with the proper manhour tables that follow, a good labor value estimate can be properly executed for any place in the world, regardless of its geographical location and whether it be today or twenty years from now.

Next, we must consider the composite rate. In order to correctly arrive at a total direct labor cost, using the manhours as appear in the following tables, this must be done.

Most organizations consider field personnel with a rating of superintendent or greater as a part of job overhead, and that of general foreman or lower as direct job labor cost. The direct manhours as appear on the following pages have been determined on this basis. Therefore, a composite rate should be used when converting the manhours to direct labor dollars.

Again, the estimator must consider labor conditions in the area where the project is to be located. He must ask himself how many men will he be allowed to use in a crew, can he use crews with mixed crafts, and how many crews of the various crafts will he need.

Following is an example that may be used to obtain a composite rate:

We assume that a certain project has a certain amount of plumbing and that we will need four (4) ten-man crews, and that only one general foreman will be needed to head the four crews:

Rate of Craft in the given area:

| | |
|---|---|
| General Foreman | $10.75 per hour |
| Foreman | 10.00 per hour |
| Journeyman | 9.50 per hour |
| Fifth Year Apprentice | 7.60 per hour |

Note: General foreman and foreman are dead weight since they do not work with their tools, however, they must be considered and charged to the composite crew.

Crew for Composite Rate:

| | | | |
|---|---|---|---|
| One General Foreman | 2 hours @ | $10.75 = | $ 21.50 |
| One Foreman | 8 hours @ | 10.00 = | 80.00 |
| Nine Journeymen | 8 hours @ | 9.50 = | 684.00 |
| Fifth Year Apprentice | 8 hours @ | 7.60 = | 60.80 |
| Total for 80 Hours | | | $846.30 |

$846.30 ÷ 80 = $10.579 Composite Manhour Rate for 100% Time

As was stated in the preface, the manhours are based on an average productivity of seventy (70) per cent for all crafts involved. Therefore, the composite rate of $10.579 as figured, becomes equal to seventy (70) per cent productivity.

Let us now assume that we have evaluated a certain project to be bid and find it to be of a low average with an overall productivity rating of only sixty-five (65) percent. This means a loss of five (5) percent of time paid per manhour. Therefore, the composite rate should have an adjustment of five (5) percent as follows.

$10.579 × 105% = $11.11 Composite Rate for 65% Productivity

Simply by multiplying the number of manhours estimated for a given block or item of work by the arrived at composite rate, a total estimated direct labor cost, in dollar value can be easily and accurately obtained.

It is hoped that the foregoing will enable the ordinary plumbing and mechanical estimator to turn out a better labor estimate by eliminating much of the guesswork.

## JOB ESTIMATING FORM

COMPANY | COMPOSITE CREW RATE | SHEET NO. ____ OF ____ | ESTIMATE NO.

PROJECT | LOCATION

DESCRIPTION OF WORK | ESTIMATOR | CHECKED BY | DATE IN | DATE DUE

| No. | Description | Unit | Quantity | Weight | | Unit Man-Hours | Total Man-Hours | Unit Labor Cost | Unit Material Cost | Total Cost | | |
|---|---|---|---|---|---|---|---|---|---|---|---|---|
| | | | | Unit | Total | | | | | Labor | Material | Total |
| | | | | | | | | | | | | |

Gulf Publishing Company, Houston Form 310

This Job Estimating Form is ideal for use when working with the Estimating Man-Hour Manuals. Prices and further information on this form available from Gulf Publishing Co., Business Forms Div., P. O. Box 2608, Houston, Texas 77001.

# CONTENTS



## Section 1 — SERVICE PIPING, PLUMBING, AND DRAINAGE

## Section 1—continued



## Section 2—DUCTWORK

## Section 2—continued



## Section 3—EQUIPMENT INSTALLATION

## Section 3—continued



## Section 4—INSULATION AND WATERPROOFING



## Section 5—INSTRUMENTS AND CONTROLS



## Section 6—ELECTRICAL INSTALLATION

## Section 6—continued



## Section 7—PAINTING



## Section 8—EARTHWORK

## Section 8—continued



## Section 9—CONCRETE AND MASONRY WORK



## Section 10—MISCELLANEOUS STEEL, FRAMING, AND SUPPORTS

# Section 11—TECHNICAL INFORMATION

# Section 1

# SERVICE PIPING, PLUMBING, AND DRAINAGE

This section covers the installation of: 1) service piping, such as water, gas, and air; 2) plumbing fixtures and components; and 3) sanitary, weather and waste drainage systems.

It is not the intent of the piping tables within this section to suffice for the installation of process systems for a structure even though the various operations may be similar in many respects.

The manhour tables that follow are average of many projects of varied nature and include operational time as may be required for a job in accordance with the notes as appear thereon.

# NOTES ON PIPE BENDS

*Minimum Bending Radii:* Manhours shown for pipe bends are based upon a minimum bending radii of 5 nominal pipe size diameters, with the exception of large sizes and/or lighter walls which must be bent on longer radii. For bends having a radius of less than 5 diameters add 50% to manhours shown.

*Welding Long Bends:* When it is necessary to weld together two or more pieces of pipe to produce the length required in the pipe bend, add the manhours for welding.

*Compound Bends:* Manhours of pipe bends other than the standard types illustrated or with bends in more than one plane are obtained by adding together the manhours of the component bends that are combined to product the compound bend.

*Bends Without Tangents:* For pipe bends (Sch. 160 and less) ordered with tangents, add 15% to manhours shown.

*Bends With Long Arcs:* For pipe bending with an arc exceeding 10 feet; add 100% to the bending manhours shown for each additional 10 feet of arc or part thereof.

*Connecting Tangents:* No. 5 Offset Bends and No. 7 U-Bends are to be considered as such only when the bends are continuous arcs or if the tangent between arcs of the same radius is 1′-0″ long or less. If the tangent between arcs is longer than 1′-0″, the bends should be considered as compound bends, *i.e.,* double angle bends, double quarter bends, etc.

No. 9 Expansion U-Bends are to be considered as such only when the bends are continuous arcs or if the tangents between the "U" and the 90° bends, of the same radius are 1′-0″ or less. When such tangents are longer than 1′-0″ the bends should be manhoured as one "U" and two 90° pipe bends. Bends from 181° to 359° should be manhoured the same as a No. 11 Bend.

*Offset Bends:* No. 5 Offset Bends are considered as such only when each angle is 90° or less and the connecting tangents between arcs are within the maximum of 1′-0″ specified in the preceding note.

*Beveled Ends:* If pipe bends are to have the ends beveled for welding add the manhours for beveling.

*Thread Ends:* If pipe bends are to have the ends threaded only, add the manhours for threading.

*Flanged Ends:* If pipe bends are to have the ends fitted with screwed flanges, slip-on weld flanges, welding neck flanges, or lap joints add the manhours applicable for this operation.

*Preparation For Intermediate Field Welds:* When pipe bends, particularly No. 9, 10, or 11, are too bulky for transporting or handling and therefore, must be furnished in two or more sections for assembly in the field, an extra charge should be made for the additional cuts and bevels.

*Unlisted Sizes:* For unlisted sizes, use the manhours of the next larger shown size.

# STANDARD TYPES OF BENDS

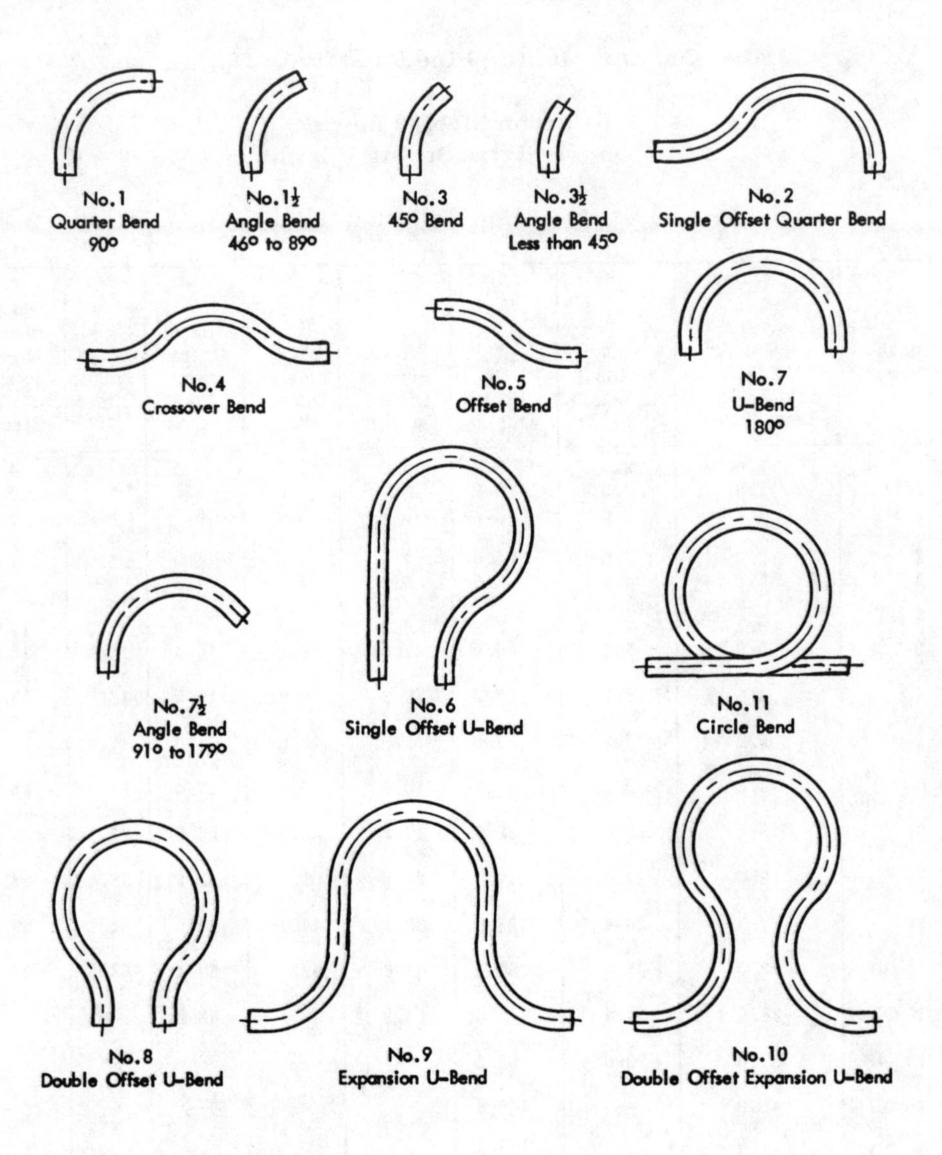

# PIPE BENDS

## Schedule Numbers 20 to 100 Inclusive

## Labor Only for Making Pipe Bends with Plain Ends

Carbon Steel Material
Double Extra Strong Weight

NET MANHOURS EACH

| Size Ins. | No.1-90° No.1-1/2- 46° to 89° No.3-45° No.3-1/2- Less than 45° | No.7 U-180° No.7-1/2- 91° to 179° | No. 5 Offset | No.2 Single Offset Quarter | No.6 Single Offset "U" | No.4 Cross-over No.9 Exp. "U" | No.8 Double Offset "U" | No.10 Double Offset Exp. No.11 Circle |
|---|---|---|---|---|---|---|---|---|
| 1 | 1.6 | 2.0 | 2.4 | 3.0 | 3.6 | 3.5 | 4.4 | 4.9 |
| 1-1/4 | 1.8 | 2.3 | 2.8 | 3.4 | 4.5 | 4.6 | 5.2 | 5.7 |
| 1-1/2 | 2.0 | 2.6 | 3.1 | 3.8 | 5.0 | 5.2 | 5.9 | 6.6 |
| 2 | 2.3 | 3.0 | 3.6 | 4.6 | 5.5 | 6.0 | 6.0 | 8.1 |
| 2-1/2 | 2.9 | 4.1 | 4.5 | 5.8 | 7.2 | 7.5 | 8.9 | 11.7 |
| 3 | 3.1 | 4.7 | 5.2 | 6.3 | 8.3 | 8.9 | 10.0 | 12.8 |
| 3-1/2 | 3.9 | 5.2 | 5.9 | 7.8 | 9.7 | 10.4 | 12.4 | 15.8 |
| 4 | 4.4 | 6.3 | 6.9 | 8.6 | 10.8 | 11.7 | 13.7 | 17.3 |
| 5 | 5.9 | 7.8 | 8.3 | 11.0 | 13.6 | 14.6 | 17.4 | 22.0 |
| 6 | 7.2 | 9.2 | 10.1 | 12.8 | 16.0 | 17.4 | 21.9 | 25.8 |
| 8 | 9.4 | 12.6 | 12.8 | 17.3 | 23.5 | 25.0 | 27.3 | 33.0 |
| 10 | 12.7 | 17.4 | 17.4 | 22.8 | 31.7 | 33.8 | 38.4 | 43.9 |
| 12 | 17.3 | 24.4 | 23.6 | 32.1 | 44.8 | 45.6 | 61.1 | 69.4 |
| 14 OD | 22.0 | 35.4 | 31.4 | 42.8 | 67.1 | 67.1 | 88.6 | 103.4 |
| 16 OD | 28.7 | 54.4 | 41.3 | 56.9 | 83.2 | 83.2 | -- | -- |
| 18 OD | 37.0 | 72.0 | 52.3 | -- | -- | -- | -- | -- |
| 20 OD | 48.0 | -- | 66.8 | -- | -- | -- | -- | -- |
| 24 OD | 81.2 | -- | 103.4 | -- | -- | -- | -- | -- |

*For General Notes on Pipe Bends—See pages 2 and 3.*

# SHOP FABRICATION—PIPE

## Carbon Steel Material Through Schedule 80

MANHOURS FOR UNITS LISTED

| Pipe Size Inches | Fabricated Pipe Per Lin. Ft. Sch. Thru 80 | Cut Pipe Per Cut | Bevel Pipe Per End | Thread Pipe Incl. Cut Per End |
|---|---|---|---|---|
| 1/4 | 0.029 | 0.11 | 0.08 | 0.21 |
| 3/8 | 0.029 | 0.11 | 0.08 | 0.21 |
| 1/2 | 0.030 | 0.11 | 0.08 | 0.21 |
| 3/4 | 0.030 | 0.11 | 0.08 | 0.21 |
| 1 | 0.031 | 0.11 | 0.09 | 0.22 |
| 1-1/4 | 0.033 | 0.11 | 0.09 | 0.22 |
| 1-1/2 | 0.035 | 0.11 | 0.09 | 0.22 |
| 2 | 0.036 | 0.11 | 0.09 | 0.22 |
| 2-1/2 | 0.039 | 0.13 | 0.10 | 0.29 |
| 3 | 0.041 | 0.16 | 0.12 | 0.31 |
| 3-1/2 | 0.044 | 0.19 | 0.15 | 0.37 |
| 4 | 0.045 | 0.21 | 0.17 | 0.43 |
| 5 | 0.048 | 0.26 | 0.21 | 0.56 |
| 6 | 0.051 | 0.34 | 0.27 | 0.69 |
| 8 | 0.063 | 0.51 | 0.40 | 1.07 |
| 10 | 0.079 | 0.80 | 0.63 | 1.56 |
| 12 | 0.096 | 1.13 | 0.89 | 2.33 |
| 14 OD | 0.116 | 1.25 | 0.98 | 2.50 |
| 16 OD | 0.138 | 1.55 | 1.22 | 3.14 |
| 18 OD | 0.161 | 1.90 | 1.50 | 3.71 |
| 20 OD | 0.189 | 2.25 | 1.82 | 5.22 |
| 24 OD | 0.210 | 3.25 | 2.44 | 6.72 |

Manhours include unloading, handling, hauling, and fabrication of any length spool piece or segment of work as outlined above.

For miter cuts less than 30° add 50%, and for miter cuts 30° or greater add 100% to the manhours.

For cutting the ends of bends or trimming fittings add 50% to the manhours.

For miter bevels and bevels on the ends of bends or shop trimmed fittings add 50% to the manhours.

Threading pipe including cut manhours are for IPS pipe threads only.

For threading the ends of bends add 100% to the manhours.

Manhours do not include field erection. See respective tables for these charges.

# SHOP FABRICATION—FITTINGS

## Carbon Steel Material

## For Bends, Headers, Necks, and Straight Runs of Pipe

NET MANHOURS EACH

| Pipe Size Inches | ATTACHING FLANGES | | | |
|---|---|---|---|---|
| | Screwed Type Make-On Reface | Screwed Type Seal Weld Back Reface | Screwed Type Seal Weld Back & Front Reface | Slip-On Type Weld Back & Front |
| 2 or less | 1.0 | 1.4 | 1.7 | 1.1 |
| 2-1/2 | 1.1 | 1.5 | 1.9 | 1.3 |
| 3 | 1.2 | 1.7 | 2.1 | 1.6 |
| 4 | 1.5 | 2.0 | 2.5 | 2.1 |
| 5 | 1.6 | 2.3 | 2.9 | 2.6 |
| 6 | 1.8 | 2.7 | 3.4 | 3.1 |
| 8 | 2.1 | 3.3 | 4.1 | 4.3 |
| 10 | 2.6 | 4.2 | 5.3 | 5.3 |
| 12 | 3.1 | 4.8 | 6.0 | 6.5 |
| 14 OD | 3.8 | 5.8 | 7.2 | 7.6 |
| 16 OD | 4.6 | 7.2 | 9.0 | 8.9 |
| 18 OD | 5.5 | 8.3 | 10.4 | 10.3 |
| 20 OD | 6.5 | 9.3 | 11.6 | 12.4 |
| 24 OD | 9.3 | 13.0 | 16.2 | 15.5 |

Manhours include handling, hauling, storing, and placing flanges into fabricated spool pieces for the type and operation as outlined above.

Manhours are for use of 150-pound pressure rated flanges. If flanges with greater pressure rating used, additional time should be allowed.

Manhours do not include pipe handling in fabrication shop or erection of spool piece or flange in field. See respective tables for these charges.

# SHOP FABRICATION—WELDING

## Carbon Steel Materials

MANHOURS REQUIRED EACH

| Pipe Size Inches | Butt Welds Schedule | | 90° Welded Nozzles Schedule | | 45° Welded Nozzles Schedule | | Concentric Swaged Ends Schedule | | Eccentric Swaged Ends Schedule | |
|---|---|---|---|---|---|---|---|---|---|---|
| | 40 | 80 | 40 | 80 | 40 | 80 | 40 | 80 | 40 | 80 |
| 1 | 0.6 | 0.7 | 1.8 | 1.9 | 2.4 | 2.5 | – | – | – | – |
| 1-1/4 | 0.7 | 0.7 | 1.9 | 2.1 | 2.5 | 2.8 | – | – | – | – |
| 1-1/2 | 0.7 | 0.8 | 2.1 | 2.3 | 2.8 | 3.1 | – | – | – | – |
| 2 | 0.8 | 0.9 | 2.2 | 2.7 | 2.9 | 3.6 | 1.4 | 1.8 | 1.6 | 1.9 |
| 2-1/2 | 1.0 | 1.1 | 2.4 | 3.3 | 3.3 | 4.4 | 1.6 | 2.1 | 1.7 | 2.5 |
| 3 | 1.1 | 1.2 | 2.8 | 3.8 | 3.8 | 5.1 | 1.7 | 2.4 | 2.0 | 2.8 |
| 3-1/2 | 1.2 | 1.4 | 3.2 | 4.3 | 4.3 | 5.6 | 2.0 | 2.8 | 2.3 | 3.3 |
| 4 | 1.3 | 1.6 | 3.5 | 4.9 | 4.8 | 6.6 | 2.3 | 3.3 | 2.7 | 3.9 |
| 5 | 1.5 | 1.9 | 4.4 | 6.0 | 5.8 | 7.9 | 3.0 | 4.2 | 3.5 | 5.2 |
| 6 | 1.8 | 2.1 | 4.7 | 6.5 | 6.2 | 8.7 | 3.6 | 5.4 | 4.3 | 6.2 |
| 8 | 2.2 | 2.8 | 5.3 | 7.5 | 7.3 | 9.9 | 5.0 | 7.8 | 6.2 | 10.1 |
| 10 | 2.7 | 4.3 | 6.0 | 10.7 | 8.1 | 14.1 | 6.6 | 12.4 | 8.1 | 17.0 |
| 12 | 3.4 | 5.6 | 8.4 | 14.4 | 11.0 | 19.4 | 13.2 | 21.0 | 17.9 | 27.9 |
| 14 OD | 4.2 | 8.1 | 9.8 | 19.2 | 13.0 | 25.5 | 18.7 | 30.3 | 26.3 | 38.8 |
| 16 OD | 5.6 | 10.5 | 12.9 | 22.7 | 17.1 | 29.9 | 24.8 | 33.5 | 34.1 | 44.3 |
| 18 OD | 7.3 | 13.9 | 16.2 | 25.5 | 21.6 | 34.4 | 38.1 | – | 51.2 | – |
| 20 OD | 7.9 | 16.5 | 18.9 | 30.0 | 25.1 | 39.7 | 42.7 | – | 58.3 | – |
| 24 OD | 11.2 | 26.3 | 23.5 | 39.0 | 30.1 | 47.5 | – | – | – | – |

Manhours include in-shop welding operations for items as outlined.

For miter butt welds, add 50% to butt weld manhours.

Manhours do not include cutting or beveling of pipe. See respective tables for these time frames.

# SHOP FABRICATION— 90° COUPLING WELDS & SOCKET WELDS

## Labor for Cutting and Welding

## Carbon Steel Material

NET MANHOURS EACH

| Pipe Sizes Inches | 90°—3000# Coupling Weld | 90°—6000# Coupling Weld | SOCKET WELDS | |
|---|---|---|---|---|
| | | | Sch. 40 & 80 Pipe | Sch. 100 & Heavier Pipe |
| 1/2" or Less | 1.4 | 1.7 | 0.5 | 0.5 |
| 3/4 | 1.6 | 1.9 | 0.5 | 0.6 |
| 1 | 1.8 | 2.2 | 0.6 | 0.7 |
| 1-1/4 | 2.1 | 2.5 | 0.8 | 0.9 |
| 1-1/2 | 2.3 | 2.8 | 0.8 | 1.0 |
| 2 | 2.9 | 3.6 | 0.9 | 1.3 |
| 2-1/2 | 3.4 | 4.2 | 1.1 | 1.4 |
| 3 | 4.0 | 4.9 | 1.2 | 1.7 |

Manhours shown are for welding of coupling to the O.D. of the pipe only.

If couplings are to be welded to the I.D. of the pipe, add 50% to the above manhours for pipe thickness up to 1 inch, and an additional 12% for each 1/4 inch or fraction thereof of pipe thickness over 1 inch.

Any coupling welded to pipe heavier than Schedule 160 should be manhoured as a 6,000-pound coupling.

For couplings welded at angles from 45° to less than 90° and couplings attached to fittings increase above manhours 50%.

For couplings welded at angles less than 45° increase above manhours 75%.

Socket welds do not include cut. See respective manhour table for this charge.

# SHOP FABRICATION—'OLET TYPE WELDS

## Labor for Cutting and Welding

## Carbon Steel Material

NET MANHOURS EACH

| NOMINAL PIPE SIZE | | Standard Weight And 2000# | Extra Strong And 3000# | Greater Than Extra Strong And 6000# |
|---|---|---|---|---|
| Outlet | Header | | | |
| 1/2 | All Sizes | 1.3 | 1.7 | 2.2 |
| 3/4 | All Sizes | 1.6 | 1.9 | 2.6 |
| 1 | All Sizes | 1.8 | 2.2 | 2.9 |
| 1-1/4 | All Sizes | 2.0 | 2.5 | 3.3 |
| 1-1/2 | All Sizes | 2.5 | 3.2 | 4.3 |
| 2 | All Sizes | 3.4 | 4.2 | 5.6 |
| 2-1/2 | All Sizes | 4.0 | 5.1 | 6.7 |
| 3 | All Sizes | 4.6 | 5.9 | 9.2 |
| 4 | All Sizes | 6.1 | 7.4 | 9.8 |
| 5 | All Sizes | 6.9 | 8.1 | 11.9 |
| 6 | All Sizes | 7.6 | 8.6 | 13.9 |
| 8 | All Sizes | 8.4 | 9.2 | 16.4 |
| 10 | All Sizes | 11.8 | 16.9 | 26.3 |
| 12 | All Sizes | 16.5 | 19.6 | 38.9 |
| 14 | 14″ and 16″ | 20.7 | 23.0 | 46.9 |
| 14 | 18″ And Larger | 18.4 | 20.7 | 51.0 |
| 16 | 16″ and 18″ | 24.7 | 26.4 | 61.2 |
| 16 | 20″ and Larger | 21.8 | 23.8 | 66.3 |
| 18 | 18″ and 20″ | 29.3 | 32.1 | 79.1 |
| 18 | 24″ and Larger | 25.8 | 28.4 | 85.2 |
| 20 | 20″ and 24″ | 35.6 | 39.0 | 87.8 |
| 20 | 26″ and Larger | 31.0 | 34.7 | 94.6 |
| 24 | 24″ and 26″ | 54.5 | 63.7 | 105.3 |
| 24 | 28″ and Larger | 45.9 | 55.1 | 113.5 |

Manhours are based on the outlet size and schedule except when the run schedule is greater than the outlet schedule, in which case the manhours are based on the outlet size and run schedule.

For elbolet or latrolet welds, and weldolets, threadolets, etc., that are attached to fittings or welded at any angle other than 90°, add 50% to the above manhours.

For sweepolet attachment welds, add 150% to the above manhours.

# FIELD FABRICATION & ERECTION—PIPE

## Carbon Steel Material Through Schedule 80

MANHOURS PER UNITS LISTED

| Pipe Size Inches | Field Erect Shop Fabricated Pipe Per Linear Foot | Field Erect Straight Run Pipe Per Linear Foot | Flame Cut Pipe Per Cut | Flame Bevel Pipe Per End | Thread Pipe Including Cut Per End |
|---|---|---|---|---|---|
| 1/4 | 0.26 | 0.16 | 0.10 | 0.06 | 0.20 |
| 3/8 | 0.27 | 0.16 | 0.10 | 0.06 | 0.20 |
| 1/2 | 0.27 | 0.16 | 0.10 | 0.06 | 0.20 |
| 3/4 | 0.28 | 0.17 | 0.10 | 0.06 | 0.20 |
| 1 | 0.29 | 0.17 | 0.10 | 0.06 | 0.20 |
| 1-1/4 | 0.30 | 0.18 | 0.12 | 0.08 | 0.20 |
| 1-1/2 | 0.32 | 0.19 | 0.12 | 0.08 | 0.20 |
| 2 | 0.34 | 0.20 | 0.12 | 0.08 | 0.20 |
| 2-1/2 | 0.36 | 0.21 | 0.12 | 0.09 | 0.27 |
| 3 | 0.39 | 0.23 | 0.15 | 0.12 | 0.30 |
| 3-1/2 | 0.40 | 0.24 | 0.18 | 0.14 | 0.36 |
| 4 | 0.41 | 0.25 | 0.21 | 0.16 | 0.42 |
| 5 | 0.44 | 0.26 | 0.24 | 0.20 | 0.54 |
| 6 | 0.47 | 0.28 | 0.33 | 0.26 | 0.70 |
| 8 | 0.57 | 0.34 | 0.46 | 0.37 | 0.97 |
| 10 | 0.72 | 0.43 | 0.64 | 0.51 | 1.38 |
| 12 | 0.88 | 0.52 | 0.86 | 0.68 | 2.07 |
| 14 OD | 1.01 | 0.64 | 1.15 | 0.91 | 2.67 |
| 16 OD | 1.27 | 0.75 | 1.61 | 1.27 | 3.45 |
| 18 OD | 1.48 | 0.88 | 2.01 | 1.58 | 4.14 |
| 20 OD | 1.74 | 1.03 | 2.48 | 1.95 | 5.68 |
| 24 OD | 1.94 | 1.15 | 3.66 | 2.88 | 7.67 |

Pipe erection manhours include job handling, hauling, rigging in place, and aligning of pipe. Prefabricated pipe manhours are for any length spool piece or segment of work.

For miter cuts and bevels less than 30°, add 50% to the manhours. For miter cuts or bevels 30° or greater add 100% to the manhours.

For cutting or beveling the ends of bends or trimmed fittings, add 50% to the manhours.

For threading the ends of bends, add 100% to the manhours.

Threading pipe manhours are for IPS pipe threads only.

# FIELD FABRICATION & ERECTION—FITTINGS

## Carbon Steel Material for Bends, Headers, Necks, and Straight Runs of Pipe

NET MANHOURS EACH

| Pipe Size Inches | ATTACHING FLANGES | | | |
|---|---|---|---|---|
| | Screwed Type Make-On Reface | Screwed Type Seal Weld Back Reface | Screwed Type Seal Weld Back & Front Reface | Slip-On Type Weld Back & Front |
| 2 or less | 1.2 | 1.6 | 2.0 | 1.3 |
| 2-1/2 | 1.3 | 1.7 | 2.2 | 1.5 |
| 3 | 1.4 | 2.0 | 2.4 | 1.8 |
| 4 | 1.7 | 2.3 | 2.9 | 2.4 |
| 5 | 1.8 | 2.6 | 3.3 | 3.0 |
| 6 | 2.1 | 3.1 | 3.9 | 3.6 |
| 8 | 2.5 | 3.9 | 4.8 | 5.1 |
| 10 | 3.1 | 5.0 | 6.2 | 6.3 |
| 12 | 3.7 | 5.7 | 7.1 | 7.7 |
| 14 OD | 4.5 | 6.8 | 8.5 | 9.0 |
| 16 OD | 5.4 | 8.5 | 10.6 | 10.5 |
| 18 OD | 6.5 | 9.9 | 12.3 | 12.2 |
| 20 OD | 7.7 | 11.0 | 13.7 | 14.6 |
| 24 OD | 11.0 | 15.3 | 19.1 | 18.3 |

Manhours include handling and installing 150-pound pressure rated flanges with the type operation as outlined above.

If flange with pressure rating greater than 150 pounds is to be used, additional time should be allowed.

Manhours do not include shop fabrication. See respective table for this charge.

# FIELD FABRICATION & ERECTION— BOLT-UPS, MAKE-ONS, & HANDLE VALVES

MANHOURS PER UNITS LISTED

| Pipe Size Inches | Bolt-Ups Each | Make-Ons Each | Handle Valves Each |
|---|---|---|---|
| 3/4 or less | 0.7 | 0.1 | 0.3 |
| 1 | 0.7 | 0.2 | 0.3 |
| 1-1/4 | 0.7 | 0.2 | 0.3 |
| 1-1/2 | 0.7 | 0.3 | 0.4 |
| 2 | 0.7 | 0.3 | 0.8 |
| 2-1/2 | 0.8 | 0.4 | 1.1 |
| 3 | 0.8 | 0.4 | 1.5 |
| 4 | 1.2 | 0.5 | 2.0 |
| 5 | 1.3 | – | 2.4 |
| 6 | 1.5 | – | 2.7 |
| 8 | 2.1 | – | 3.4 |
| 10 | 2.7 | – | 4.2 |
| 12 | 3.4 | – | 5.1 |
| 14 OD | 3.8 | – | 6.0 |
| 16 OD | 4.4 | – | 7.1 |
| 18 OD | 4.8 | – | 8.1 |
| 20 OD | 5.5 | – | 9.2 |
| 24 OD | 6.6 | – | 10.3 |

Manhours include time necessary to complete the above-outlined operations in the field.

*Make-On* manhours do not include backwelding. This must be added if necessary.

*Handle Valves* manhours include handling and positioning valves only and do not include make-ons or bolt-ups. These must be added.

*Bolt-Up and Handle Valve* manhours are for pressure ratings of up to 150 pounds.

Manhours do not include shop work. See respective tables for these charges.

# FIELD FABRICATION & ERECTION—WELDING

## Carbon Steel Materials

MANHOURS REQUIRED EACH

| Pipe Size Inches | Butt Welds Schedule | | 90° Welded Nozzles Schedule | | 45° Welded Nozzles Schedule | | Concentric Swaged Ends Schedule | | Eccentric Swaged Ends Schedule | |
|---|---|---|---|---|---|---|---|---|---|---|
| | 40 | 80 | 40 | 80 | 40 | 80 | 40 | 80 | 40 | 80 |
| 1 | 0.7 | 0.8 | 2.1 | 2.2 | 2.8 | 2.9 | – | – | – | – |
| 1-1/4 | 0.8 | 0.8 | 2.2 | 2.4 | 2.9 | 3.2 | – | – | – | – |
| 1-1/2 | 0.8 | 0.9 | 2.4 | 2.6 | 3.2 | 3.6 | – | – | – | – |
| 2 | 1.0 | 1.0 | 2.5 | 3.1 | 3.3 | 4.1 | 1.6 | 2.1 | 1.8 | 2.2 |
| 2-1/2 | 1.2 | 1.3 | 2.8 | 3.8 | 3.8 | 5.1 | 1.8 | 2.4 | 2.0 | 2.9 |
| 3 | 1.3 | 1.4 | 3.2 | 4.4 | 4.4 | 5.9 | 2.0 | 2.8 | 2.3 | 3.2 |
| 3-1/2 | 1.4 | 1.6 | 3.7 | 4.9 | 4.9 | 6.4 | 2.3 | 3.2 | 2.6 | 3.8 |
| 4 | 1.5 | 1.8 | 4.0 | 5.6 | 5.5 | 7.6 | 2.6 | 3.8 | 3.1 | 4.5 |
| 5 | 1.7 | 2.1 | 5.1 | 6.9 | 6.7 | 9.1 | 3.5 | 4.8 | 4.0 | 6.0 |
| 6 | 2.0 | 2.5 | 5.4 | 7.5 | 7.1 | 10.0 | 4.1 | 6.2 | 4.9 | 7.1 |
| 8 | 2.6 | 3.3 | 6.3 | 8.9 | 8.6 | 11.7 | 5.9 | 9.2 | 7.3 | 11.9 |
| 10 | 3.1 | 5.1 | 7.1 | 12.6 | 9.6 | 16.6 | 7.8 | 14.6 | 10.0 | 20.1 |
| 12 | 4.1 | 6.6 | 9.9 | 17.0 | 13.0 | 22.9 | 15.6 | 24.8 | 21.1 | 32.9 |
| 14 OD | 5.0 | 9.6 | 11.6 | 22.7 | 15.3 | 30.1 | 22.1 | 35.8 | 31.0 | 45.8 |
| 16 OD | 6.6 | 12.4 | 15.2 | 26.8 | 20.2 | 35.3 | 29.3 | 39.5 | 40.2 | 52.3 |
| 18 OD | 8.6 | 16.4 | 19.1 | 30.1 | 25.5 | 40.6 | 45.0 | – | 60.4 | – |
| 20 OD | 9.4 | 19.5 | 22.3 | 35.4 | 29.6 | 46.9 | 50.4 | – | 68.8 | – |
| 24 OD | 13.3 | 25.2 | 27.7 | 46.0 | 35.5 | 55.0 | – | – | – | – |

Manhours include field welding of items at erection location.

For miter butt welds, add 50% to manhours.

Manhours do not include cutting or beveling of pipe. See respective tables for these time requirements.

# FIELD ERECTION—
# 90° COUPLING WELDS & SOCKET WELDS

## Labor for Cutting and Welding
## Carbon Steel Material

NET MANHOURS EACH

| Pipe Sizes Inches | 90°—3000# Coupling Weld | 90°—6000# Coupling Weld | SOCKET WELDS | |
|---|---|---|---|---|
| | | | Sch. 40 & 80 Pipe | Sch. 100 & Heavier Pipe |
| 1/2″ or Less | 1.4 | 1.7 | 0.5 | 0.5 |
| 3/4 | 1.6 | 1.9 | 0.5 | 0.6 |
| 1 | 1.8 | 2.2 | 0.6 | 0.7 |
| 1-1/4 | 2.1 | 2.5 | 0.8 | 0.9 |
| 1-1/2 | 2.3 | 2.8 | 0.8 | 1.0 |
| 2 | 2.9 | 3.6 | 0.9 | 1.3 |
| 2-1/2 | 3.4 | 4.2 | 1.1 | 1.4 |
| 3 | 4.0 | 4.9 | 1.2 | 1.7 |

Manhours shown are for welding of coupling to the O.D. of the pipe only.

If couplings are to be welded to the I.D. of the pipe, add 50% to the manhours. For pipe thickness up to 1 inch, add an additional 12% for each 1/4 inch or fraction thereof of pipe thickness over 1 inch.

Any coupling welded to pipe heavier than Schedule 160 should be manhoured as a 6000-pound coupling.

For couplings welded at angles from 45° to less than 90° and couplings attached to fittings increase above manhours 50%.

For couplings welded at angles less than 45° increase above manhours 75%.

Socket welds do not include cut. See respective manhour table for this charge.

# FIELD ERECTION—'OLET TYPE WELDS

## Labor Cutting and Welding

## Carbon Steel Material

NET MANHOURS EACH

| NOMINAL PIPE SIZE | | Standard Weight And 2000# | Extra Strong And 3000# | Greater Than Extra Strong And 6000# |
|---|---|---|---|---|
| Outlet | Header | | | |
| 1/2 | All Sizes | 1.3 | 1.7 | 2.2 |
| 3/4 | All Sizes | 1.6 | 1.9 | 2.6 |
| 1 | All Sizes | 1.8 | 2.2 | 2.9 |
| 1-1/4 | All Sizes | 2.0 | 2.5 | 3.3 |
| 1-1/2 | All Sizes | 2.5 | 3.2 | 4.3 |
| 2 | All Sizes | 3.4 | 4.2 | 5.6 |
| 2-1/2 | All Sizes | 4.0 | 5.1 | 6.7 |
| 3 | All Sizes | 4.6 | 5.9 | 9.2 |
| 4 | All Sizes | 6.1 | 7.4 | 9.8 |
| 5 | All Sizes | 6.9 | 8.1 | 11.9 |
| 6 | All Sizes | 7.6 | 8.6 | 13.9 |
| 8 | All Sizes | 8.4 | 9.2 | 16.4 |
| 10 | All Sizes | 11.8 | 16.9 | 26.3 |
| 12 | All Sizes | 16.5 | 19.6 | 38.9 |
| 14 | 14″ and 16″ | 20.7 | 23.0 | 46.9 |
| 14 | 18″ And Larger | 18.4 | 20.7 | 51.0 |
| 16 | 16″ and 18″ | 24.7 | 26.4 | 61.2 |
| 16 | 20″ and Larger | 21.8 | 23.8 | 66.3 |
| 18 | 18″ and 20″ | 29.3 | 32.1 | 79.1 |
| 18 | 24″ and Larger | 25.8 | 28.4 | 85.2 |
| 20 | 20″ and 24″ | 35.6 | 39.0 | 87.8 |
| 20 | 26″ and Larger | 31.0 | 34.7 | 94.6 |
| 24 | 24″ and 26″ | 54.5 | 63.7 | 105.3 |
| 24 | 28″ and Larger | 45.9 | 55.1 | 113.5 |

Manhours are based on the outlet size and schedule except when the run schedule is greater than the outlet schedule, in which case the manhours are based on the outlet size and run schedule.

For elbolet or latrolet welds and weldolets, thredolets, etc., that are attached to fittings or welded at any angle other than 90°, add 50% to the manhours.

For sweepolet attachment welds, add 150% to the manhours.

# FIELD FABRICATION & ERECTION— PIPE & FITTINGS

## Nickel or Chromium Plate Through Schedule 80

MANHOURS PER UNITS LISTED

| Size Inches | Pipe Per Lin. Ft. | Cut & Thread Each | Make-Ons Each | Bolt-Ups Each |
|---|---|---|---|---|
| ¾ or less | 0.19 | 0.11 | 0.11 | 0.8 |
| 1 | 0.19 | 0.11 | 0.22 | 0.8 |
| 1¼ | 0.20 | 0.11 | 0.22 | 0.8 |
| 1½ | 0.21 | 0.11 | 0.33 | 0.8 |
| 2 | 0.22 | 0.22 | 0.33 | 0.8 |

Manhours are for the field fabrication and erection of items outlined above, and include handling and hauling as may be necessary.

Manhours do not include scaffolding. See respective table for this time requirement.

# FIELD FABRICATION & ERECTION— PIPE & FITTINGS

## Stainless Steel Through Schedule 80

MANHOURS PER UNITS LISTED

| Pipe Size Inches | Pipe Per Linear Foot | Cut & Thread Each | Make-Ons Each | Bolt-Ups Each |
|---|---|---|---|---|
| 3/4 or less | 0.22 | 0.28 | 0.12 | 1.02 |
| 1 | 0.22 | 0.28 | 0.25 | 1.02 |
| 1-1/4 | 0.24 | 0.28 | 0.26 | 1.02 |
| 1-1/2 | 0.25 | 0.28 | 0.40 | 1.02 |
| 2 | 0.26 | 0.28 | 0.41 | 1.02 |
| 2-1/2 | 0.28 | 0.37 | 0.55 | 1.16 |
| 3 | 0.31 | 0.43 | 0.56 | 1.19 |
| 3-1/2 | 0.32 | 0.52 | 0.57 | 1.49 |
| 4 | 0.35 | 0.61 | 0.72 | 1.85 |
| 5 | 0.36 | 0.80 | 0.84 | 2.15 |
| 6 | 0.40 | 1.06 | 1.00 | 2.40 |

Manhours are for field fabrication and erection of items as outlined, and include job handling and hauling as may be necessary.

Manhours do not include scaffolding. See respective table for this time requirement.

# FIELD FABRICATION & ERECTION— TYPE L & K COPPER TUBING WITH WROUGHT COPPER FITTINGS

MANHOURS PER UNITS LISTED

| Pipe Size Inches | Pipe Per Lin. Ft. | Couplings Each | Ells Each | Tees Each | Flanges Each | Reducers Each | Adapters Each | Unions Each | Caps & Plugs Each | Valves Each |
|---|---|---|---|---|---|---|---|---|---|---|
| 1/8 | .20 | .16 | .17 | .24 | – | – | – | – | – | .35 |
| 3/8 | .20 | .16 | .17 | .24 | .16 | .18 | .15 | .18 | .08 | .35 |
| 1/2 | .20 | .22 | .23 | .33 | .22 | .25 | .20 | .25 | .11 | .40 |
| 3/4 | .21 | .28 | .29 | .42 | .28 | .32 | .25 | .35 | .14 | .45 |
| 1 | .21 | .43 | .44 | .63 | .32 | .50 | .40 | .55 | .22 | .60 |
| 1-1/4 | .22 | .80 | .85 | 1.20 | .64 | .85 | .75 | .90 | .40 | 1.00 |
| 1-1/2 | .23 | .85 | .90 | 1.26 | .70 | .90 | .80 | 1.00 | .42 | 1.15 |
| 2 | .25 | .96 | 1.00 | 1.44 | .90 | 1.00 | .92 | 1.15 | .50 | 1.25 |
| 2-1/2 | .26 | 1.50 | 1.55 | 2.25 | 1.15 | 1.65 | 1.35 | 1.85 | .75 | 2.10 |
| 3 | .28 | 1.95 | 2.00 | 2.91 | 1.40 | 2.00 | 1.80 | 2.25 | .95 | 2.50 |
| 3-1/2 | .30 | 1.97 | 2.15 | 2.96 | 1.48 | 2.08 | 1.86 | 2.48 | .97 | 2.74 |
| 4 | .31 | 2.00 | 2.25 | 3.00 | 1.65 | 2.15 | 1.90 | 2.75 | 1.00 | 3.00 |
| 5 | .39 | 2.50 | 2.80 | 3.75 | 2.06 | 2.69 | 2.38 | 3.13 | 1.25 | 3.10 |
| 6 | .47 | 2.70 | 3.12 | 4.20 | 2.40 | 3.12 | 2.76 | 3.66 | 1.38 | 3.30 |
| 8 | .62 | 3.20 | 4.00 | 5.20 | 3.04 | 4.00 | 3.60 | 4.80 | 1.68 | 4.00 |
| 10 | .78 | 3.50 | 4.50 | 6.00 | 3.50 | 4.70 | 4.00 | 5.50 | 1.80 | 4.50 |

Manhours are for job handling, hauling, field fabrication, and installing pipe, fittings, and valves as listed.

Pipe manhours are for placing pipe in trench to 7′0″ deep. For installation of pipe in other areas increase the manhours by the following percentage for the conditions listed:

Building walls through first floor –2.00
Building walls above first floor –2.50
Ceilings through first floor –1.50
Ceilings second floor and above –1.75
Interconnecting equipment inside of building–3.00

Fitting and valve installation manhours are for positioning and installing regardless of location.

Manhours do not include earthwork or scaffolding. See respective tables for these time frames.

# FIELD FABRICATION AND ERECTION— BRASS PIPE, CAST BRONZE, SCREWED FITTINGS, & BRONZE VALVES

MANHOURS REQUIRED PER FITTING, VALVE, OR LINEAR FOOT OF PIPE

| Pipe Size Inches | Pipe | Couplings | Ells | Tees | Crosses |
|---|---|---|---|---|---|
| 1/2 | .10 | .20 | .20 | .30 | .40 |
| 3/4 | .11 | .20 | .20 | .30 | .40 |
| 1 | .11 | .40 | .40 | .60 | .80 |
| 1-1/4 | .11 | .40 | .40 | .60 | .80 |
| 1-1/2 | .12 | .60 | .60 | .90 | 1.20 |
| 2 | .13 | .60 | .60 | .90 | 1.20 |
| 2-1/2 | .13 | .80 | .80 | 1.20 | 1.60 |
| 3 | .14 | .80 | .80 | 1.20 | 1.60 |
| 4 | .15 | 1.00 | 1.00 | 1.50 | 3.00 |

| Pipe Size Inches | Companion Flanges | Reducers | Unions | Caps And Plugs | Valves |
|---|---|---|---|---|---|
| 1/2 | .10 | .20 | .30 | .10 | .40 |
| 3/4 | .10 | .20 | .30 | .10 | .45 |
| 1 | .20 | .30 | .60 | .20 | .60 |
| 1-1/4 | .20 | .40 | .60 | .20 | 1.00 |
| 1-1/2 | .30 | .50 | .90 | .30 | 1.15 |
| 2 | .30 | .60 | .90 | .30 | 1.25 |
| 2-1/2 | .40 | .70 | 1.20 | .40 | 2.10 |
| 3 | .40 | .80 | 1.20 | .40 | 2.50 |
| 4 | .50 | .90 | 1.50 | .50 | 3.00 |

Manhours are for job handling, hauling, field fabrication, and installing Schedule 40 pipe and fittings and valves through 175# rating.

Pipe manhours are for placing pipe in trench to 7′0″ deep. For installation of pipe in other areas, increase the manhours by the following percentage for the conditions listed:

Building walls through first floor —2.00
Building walls above first floor —2.50
Ceilings through first floor —1.50
Ceilings second floor and above —1.75
Interconnecting equipment inside of building—3.00

Fitting and valve installation manhours are for positioning and installing regardless of location.

Manhours do not include earthwork or scaffolding. See respective tables for these time frames.

# FIELD FABRICATION & ERECTION—HI-TEMP CPVC-PLASTIC PIPE

MANHOURS PER UNITS LISTED

| Pipe Size Inches | Handle Pipe Per L.F. | Cemented Socket Joints–Ea. | Saddles Each | Handle Valves CPCV Body–Ea. |
|---|---|---|---|---|
| 1/2 | .07 | .20 | .38 | .13 |
| 3/4 | .07 | .22 | .39 | .16 |
| 1 | .07 | .25 | .40 | .17 |
| 1-1/4 | .08 | .27 | .43 | .20 |
| 1-1/2 | .08 | .29 | .45 | .25 |
| 2 | .09 | .33 | .50 | .35 |
| 2-1/2 | .09 | .38 | .55 | .58 |
| 3 | .10 | .45 | .63 | .85 |
| 4 | .11 | .55 | .73 | 1.25 |
| 6 | .12 | .70 | .90 | 1.60 |
| 8 | .14 | .80 | 1.00 | 1.95 |
| 10 | .17 | 1.00 | 1.20 | 2.50 |
| 12 | .20 | 1.25 | 1.40 | 3.00 |

*Handle Pipe* manhours include handling, hauling, rigging, and aligning in place.

*Cement Socket Joint* manhours include cut, square, ream, fit-up, and make joint.

*Saddle* manhours include fit-up, drill hole in header and cement saddle to header. Maximum hole size is assumed to be 1-1/2 inch. For larger branch lines the use of tees should be estimated. The size of the header not the size of the saddle determines the manhours that apply.

*Handle Valve* manhours include handling, hauling, and positioning of valve only. Connections of the type as required must be added.

Units are for all wall thickness.

Units do not include scaffolding. See respective table for this charge.

Manhours are for placing pipe underground to 7′0″ deep. For installation of pipe in other areas, increase the manhours by the following percentage for the conditions listed:

In building walls through first floor—2.50
In building walls above first floor —3.00
In ceilings through first floor —1.70
In ceilings above first floor —2.25

# FUSEAL FLAME RETARDANT POLYPROPYLENE PIPE

MANHOURS PER UNITS LISTED

| Pipe Size Inches | HANDLE PIPE PER LINEAR FOOT | | | | |
|---|---|---|---|---|---|
| | In Trench to 7'0" Deep | In Bldg. Walls Thru First Floor | In Bldg. Walls Above First Floor | In Ceilings Thru First Floor | In Ceilings Above First Floor |
| 1-1/2 | .08 | .20 | .24 | .14 | .18 |
| 2 | .09 | .23 | .27 | .15 | .20 |
| 3 | .10 | .25 | .30 | .17 | .23 |
| 4 | .11 | .28 | .33 | .19 | .25 |

| Pipe Size Inches | MAKE-UPS EACH | | MISCELLANEOUS ITEMS EACH | | |
|---|---|---|---|---|---|
| | Electric Fused | Screwed | "P" Trap | Jar Trap | Jar with Gasket |
| 1-1/2 | .17 | .30 | .40 | .70 | .25 |
| 2 | .20 | .30 | .50 | .70 | .25 |
| 3 | .27 | .40 | .70 | .70 | .25 |
| 4 | .33 | .50 | .90 | .70 | .25 |

| Pipe Size Inches | MISCELLANEOUS ITEMS EACH | | | | |
|---|---|---|---|---|---|
| | Drum Trap | Flat or Domed Cap | Sink Strainer Assembly | Cup Sinks— Oval | Wall Brackets |
| 1-1/2 | .70 | .30 | .90 | 6"x3"—1.20 | .10 |
| 2 | .70 | .30 | — | — | .10 |
| 3 | — | — | — | — | .10 |
| 4 | — | — | — | — | .10 |

*Handle Pipe* manhours include job handling, hauling, and aligning in place.

*Make-Up* manhours include the handling of the fitting and joint make-up as outlined. Ells = two make-ups, tees = three make-ups, etc.

*Miscellaneous Item* manhours include the handling, setting, aligning, and complete make-up of the items as outlined.

Units do not include scaffolding. See respective table for this charge.

# FIELD FABRICATION & ERECTION— SCHEDULE 40 CEMENT LINED CARBON STEEL PIPE WITH STANDARD FITTINGS

MANHOURS PER UNITS LISTED

| Pipe Size Ins. | Handle Pipe Per Foot | Cutting Pipe Per Cut | Butt Welds Each | Sleeve Joint With Two Welds Each | 90° Welded Nozzle Each | Smooth On Cement Per Joint |
|---|---|---|---|---|---|---|
| 6 | .40 | .65 | 2.0 | 3.30 | 6.05 | .35 |
| 8 | .50 | 1.00 | 2.6 | 4.60 | 7.30 | .50 |
| 10 | .65 | 1.20 | 3.1 | 5.70 | 8.30 | .60 |
| 12 | .80 | 1.45 | 4.1 | 6.90 | 11.35 | .65 |
| 14 | .95 | 2.30 | 5.0 | 7.90 | 13.90 | .80 |
| 16 | 1.25 | 2.95 | 6.6 | 9.20 | 18.15 | .95 |
| 18 | 1.40 | 3.70 | 8.6 | 10.40 | 22.80 | 1.10 |
| 20 | 1.75 | 4.65 | 9.4 | 12.40 | 26.95 | 1.25 |
| 24 | 2.20 | 5.90 | 13.3 | 15.50 | 33.60 | 1.50 |

*Handle Pipe Units:* Manhours include handling, unloading, hauling to storage and erection site, setting, and aligning.

*Cutting Pipe Units:* Manhours include cutting pipe and lining. Lining to be cut square and flush with ends.

*Butt Weld Units:* Manhours include making complete electric weld. Cement lining should be wet with water around welding area.

*Sleeve Joint Units:* Manhours include slipping on of sleeve, aligning, and welding at both ends.

*90° Welded Nozzle Units:* Manhours include complete operations for welding nozzle.

*Smooth on Cement Units:* Manhours include mixing and patching weld joints with cement.
Manhours do not include excavation or racks or supports. See respective tables for these charges.

# FIELD FABRICATION & ERECTION— OVERHEAD TRANSITE PRESSURE PIPE—CLASS 150

NET MANHOURS PER UNITS LISTED

| Size Inches | Pipe Per L. F. | Make-Ons Each | Bolt-Ups Each |
|---|---|---|---|
| 4 | .20 | .50 | 1.2 |
| 6 | .25 | .75 | 1.5 |
| 8 | .30 | .85 | 2.1 |
| 10 | .40 | 1.00 | 2.7 |
| 12 | .50 | 1.25 | 3.4 |
| 14 | .60 | 1.50 | 3.8 |
| 16 | .70 | 1.75 | 4.4 |

*Pipe Units:* Pipe units include rigging, erecting, and aligning of pipe.

*Make-On Units:* Make-on units include erecting, aligning pouring joint. Ells = Two make-ons, tees = three make-ons, etc.

*Bolt-Up Units:* Bolt-up units include bolting together of flanged connections.

*All Units:* All units include unloading, handling, and hauling to storage and erection site.

Manhours do not include support or scaffolding. See respective tables for these charges.

Transite pipe = 4 inches I.D. and above. It is usually supplied in standard 13-foot lengths.

# CAST IRON SOIL PIPE

MANHOURS REQUIRED

| MANHOURS PER FOOT | | MANHOURS PER MAKE-ON | | | |
|---|---|---|---|---|---|
| Pipe Size Inches | Pipe Set and Align | Lead Joint | Cement Joint | Bituminous Joint | Rubber Slip Joing |
| 2 | 0.08 | 0.40 | 0.26 | 0.20 | 0.18 |
| 3 | 0.12 | 0.44 | 0.30 | 0.22 | 0.20 |
| 4 | 0.14 | 0.50 | 0.35 | 0.25 | 0.23 |
| 5 | 0.15 | 0.54 | 0.36 | 0.27 | 0.24 |
| 6 | 0.18 | 0.57 | 0.37 | 0.29 | 0.26 |
| 8 | 0.23 | 0.70 | 0.50 | 0.35 | 0.32 |
| 10 | 0.30 | 0.88 | 0.63 | 0.44 | 0.40 |
| 12 | 0.36 | 1.05 | 0.75 | 0.52 | 0.47 |
| 15 | 0.45 | 1.31 | 0.94 | 0.66 | 0.60 |

Manhours include handling, hauling, setting and aligning in trench, and make-up of joint as outlined above.

Manhours do not include excavation or backfill. See respective tables for these charges.

# UNDERGROUND VITRIFIED CLAY & CONCRETE PIPE

LABOR IN MANHOURS

| Size Inches | CONCRETE PIPE (Not Reinforced) | | VITRIFIED CLAY PIPE | |
|---|---|---|---|---|
| | Set & Align Pipe Per Foot | Cement Poured Joint Each | Set & Align Pipe Per Foot | Poured Joint Each |
| 4 | 0.07 | 0.20 | 0.07 | 0.25 |
| 6 | 0.08 | 0.25 | 0.07 | 0.29 |
| 8 | 0.10 | 0.32 | 0.07 | 0.35 |
| 10 | 0.11 | 0.39 | 0.08 | 0.43 |
| 12 | 0.15 | 0.50 | 0.10 | 0.62 |
| 15 | 0.19 | 0.75 | 0.11 | 0.89 |
| 18 | 0.28 | 0.95 | 0.14 | 1.14 |
| 21 | 0.29 | 1.15 | 0.19 | 1.38 |
| 24 | 0.32 | 1.25 | 0.25 | 1.63 |
| 30 | 0.40 | 1.56 | 0.31 | 2.04 |
| 36 | 0.48 | 1.88 | 0.37 | 2.44 |
| 42 | 0.56 | 2.19 | 0.44 | 2.85 |
| 48 | 0.64 | 2.50 | 0.50 | 3.26 |
| 60 | 0.80 | 3.13 | 0.62 | 4.07 |

Manhours includes handling, hauling, setting in trench, and aligning. Manhours for joint or connection of fittings is for one make-on only.

No labor for excavation or backfill is included. Add from respective pages for these charges.

For reinforced concrete pipe add 5% to manhours listed for concrete pipe.

# UNDERGROUND 150 LBS. B & S CAST IRON PIPE

LABOR IN MANHOURS

| MAN HOURS PER FOOT | | PER MAKE-ON | | |
|---|---|---|---|---|
| | | 150 Lb. B & S Fittings | | |
| Size Inches | Pipe Set & Align | Lead & Mech. Joint | Cement Joint | Sulphur Joint |
| 4 | 0. 09 | 0. 50 | 0. 35 | 0. 25 |
| 6 | 0. 11 | 0. 57 | 0. 37 | 0. 29 |
| 8 | 0. 14 | 0. 70 | 0. 50 | 0. 35 |
| 10 | 0. 17 | 0. 85 | 0. 60 | 0. 43 |
| 12 | 0. 24 | 1. 23 | 0. 95 | 0. 62 |
| 14 | 0. 35 | 1. 78 | 1. 25 | 0. 89 |
| 16 | 0. 45 | 2. 28 | 1. 60 | 1. 14 |
| 18 | 0.53 | 2.68 | 1.89 | 1.34 |
| 20 | 0.63 | 3.19 | 2.24 | 1.59 |
| 24 | 0.79 | 4.00 | 2.81 | 2.00 |

Pipe manhours include handling, hauling, setting and aligning in trench.

Fitting manhours includes one make-on.

Manhours must be added for excavation. See respective pages for this charge.

# SOCKET CLAMPS FOR CAST IRON PIPE

NET LABOR IN MANHOURS

| Pipe Size Inches | Friction Clamps Complete | Positive Clamps Complete |
|---|---|---|
| 4 | 0.25 | 0.30 |
| 6 | 0.28 | 0.33 |
| 8 | 0.33 | 0.38 |
| 10 | 0.38 | 0.43 |
| 12 | 0.45 | 0.52 |
| 14 | 0.52 | 0.62 |
| 16 | 0.60 | 0.75 |
| 18 | 0.68 | 0.85 |
| 20 | 0.75 | 0.95 |
| 24 | 0.88 | 1.10 |

Manhours are for labor only and include handling, hauling, and complete installation in all cases.

# PIPE COATED WITH TAR & FIELD WRAPPED BY MACHINE

NET MANHOURS PER LINEAR FOOT

| Nominal Pipe Size | Man Hours Per Foot | Nominal Pipe Size | Man Hours Per Foot |
|---|---|---|---|
| 3/4 | 0.04 | 22 | 0.50 |
| 1 | 0.04 | 24 | 0.54 |
| 1-1/4 | 0.05 | 26 | 0.59 |
| 1-1/2 | 0.06 | 28 | 0.63 |
| 2 | 0.07 | 30 | 0.68 |
| 2-1/2 | 0.08 | 32 | 0.73 |
| 3 | 0.09 | 34 | 0.78 |
| 4 | 0.12 | 36 | 0.82 |
| 5 | 0.13 | 38 | 0.87 |
| 6 | 0.16 | 40 | 0.91 |
| 8 | 0.20 | 42 | 0.96 |
| 10 | 0.25 | 44 | 1.00 |
| 12 | 0.28 | 46 | 1.05 |
| 14 | 0.32 | 48 | 1.10 |
| 16 | 0.37 | 54 | 1.24 |
| 18 | 0.41 | 60 | 1.38 |
| 20 | 0.45 | -- | -- |

Manhours include:

- Sandblast commercially
- Apply one prime coat of pipeline primer
- Apply 3/32″ pipeline enamel
- Apply two ply of 15# tarred felt
- Apply one seal coat of pipeline enamel

For hand coating and wrapping add 100% to above manhours.

# HANGERS & SUPPORTS

FIELD ERECTION MANHOURS

| Type of Hanger | Hanger Suspended From | Man Hours Per Hanger* | | | |
|---|---|---|---|---|---|
| | | Hanger Fastened To | | | |
| | | Steel | Concrete or Masonry | Wood | Existing Pipe |
| PATENT | Welded Clip Angle | 1.50 | -- | -- | -- |
| | Clip Angle — Ramset | 1.00 | -- | -- | -- |
| | Female Stud or Male Stud & Coupling — Ramset | .60 | .60 | -- | -- |
| Clevis Hanger Band Hanger Ring Hanger Expansion Hanger | Female Stud or Male Stud & Coupling — Nelson Stud Welder | .60 | -- | -- | -- |
| | Beam Clamp or Corn Clamp | 1.30 | -- | -- | -- |
| | Cinch Anchor | -- | 2.00 | -- | -- |
| | Bolt or Strap | -- | -- | 1.60 | -- |
| | Band and Rod | -- | -- | -- | 1.00 |
| PATENT | Welded Clip Angles | 2.00 | -- | -- | -- |
| | Clip Angle — Ramset | 1.50 | -- | -- | -- |
| | Female Stud or Male Stud & Coupling — Ramset | 1.20 | -- | -- | -- |
| Trapeze Hanger (1' - 4' Bar) | Female Stud or Male Stud & Coupling — Nelson Stud Welder | 1.20 | -- | -- | -- |
| | Beam Clamp or Corn Clamp | 2.00 | -- | -- | -- |
| | Cinch Anchor | -- | 4.00 | -- | -- |

**The patent hanger allowances are for supporting pipe through 4″ size.*

*Fabricated Hangers (Angles, Channels, Etc.):* 0.08 manhours per pound with a minimum time of 1 manhour regardless of weight.

The following factors should be applied for sizes over 4″:

6″ – 1.20 manhours
8″ – 1.50 manhours
10″ – 1.80 manhours
12″ – 2.20 manhours

*Fabrication:* Labor only for fabrication of other than standard manufactured hangers and supports can be performed at 0.07 manhours per pound.

# INSTALL STEEL SEPTIC TANKS

NET MANHOURS EACH

| Capacity | Manhours |
|---|---|
| 200 Gallons | 5.0 |
| 300 Gallons | 6.0 |
| 500 Gallons | 8.0 |
| 1000 Gallons | 10.5 |

Manhours include procuring, handling, hauling, setting and aligning septic tanks as outlined above.

Manhours do not include excavation or backfill. See respective tables for these charges.

See respective table for concrete septic tanks.

# ROUGH-IN BATHROOM FIXTURES

MANHOURS EACH

| Item | Manhours | | |
|---|---|---|---|
| | Plumber | Helper | Total |
| Lavatory | 6.30 | 6.30 | 12.60 |
| Water closet | 7.20 | 7.20 | 14.40 |
| Urinal with stall | 4.68 | 4.68 | 9.36 |
| Urinal – pedestal type | 4.50 | 4.50 | 9.00 |
| Shower with stall | 7.20 | 7.20 | 14.40 |
| Five person shower group | 14.80 | 14.80 | 29.60 |
| Shower head and mixer | 5.20 | 5.20 | 10.40 |
| Safety shower | 2.40 | 2.40 | 4.80 |
| Bath tub | 5.40 | 5.40 | 10.80 |
| Slop sink | 4.00 | 4.00 | 8.00 |

Manhours include handling, hauling, fabricating and installing all roughing in material for drainage, hot and cold water supply, vent piping and fixture supports within a 10-foot radius of fixture location.

Manhours do not include earthwork or fixture installation. See respective tables for these charges.

# ROUGH-IN MISCELLANEOUS FIXTURES AND DRAINS

MANHOURS EACH

| Item | Manhours | | |
|---|---|---|---|
| | Plumber | Helper | Total |
| 2-Inch Shower Drain | 3.20 | 3.20 | 6.40 |
| 4-Inch Cast Iron Arca Drain | 3.20 | 3.20 | 6.40 |
| 4-Inch Cast Iron Floor Drain with Trap | 4.80 | 4.80 | 9.60 |
| 4-Inch Roof Drain with Expansion-Joint | 9.60 | 9.60 | 19.20 |
| 6-Person Wash Fountain, 54-Inch Semicircle | 13.60 | 13.60 | 27.20 |
| 4-Person Wash Fountain, 36-Inch Semicircle | 9.60 | 9.60 | 19.20 |
| 6-Person Wash Fountain, 36-Inch Round | 12.80 | 12.80 | 25.60 |
| 10-Person Wash Fountain, 54-Inch Round | 16.80 | 16.80 | 33.60 |
| Drinking Fountain | 2.40 | 2.40 | 4.80 |
| Eyewash | 2.00 | 2.00 | 4.00 |

Manhours include handling, hauling, fabricating and installing all roughing-in material for drainage, hot and cold water supply and fixture supports within a 10-feet radius of fixture location.

Manhours do not include earthwork or fixture installation. See respective tables for these charges.

# FINISHED INSTALLATION – BATHROOM FIXTURES

MANHOURS EACH

| Item | MANHOURS | | |
|---|---|---|---|
| | Plumber | Helper | Total |
| Lavatory | 1.60 | 1.60 | 3.20 |
| Water Closet | 3.00 | 3.00 | 6.00 |
| Urinal with Stall | 6.00 | 6.00 | 12.00 |
| Urinal – Pedestal Type | 4.00 | 4.00 | 8.00 |
| Shower with Stall | 2.40 | 2.40 | 4.80 |
| Five Person Shower Group | 8.40 | 8.40 | 16.80 |
| Shower Head and Mixer | 1.20 | 1.20 | 2.40 |
| Safety Shower | 1.60 | 1.60 | 3.20 |
| Bath Tub | 6.00 | 6.00 | 12.00 |
| Slop Sink | 3.04 | 3.04 | 6.08 |
| Hot Water Heater – Gas Automatic | 3.40 | 3.40 | 6.80 |

Manhours include the handling, hauling and installing of fixtures and fixture fittings.

Manhours do not include earthwork or roughing in. See respective tables for these charges.

# FINISHED INSTALLATION – MISCELLANEOUS FIXTURES AND DRAINS

MANHOURS EACH

| Item | MANHOURS | | |
|---|---|---|---|
| | Plumber | Helper | Total |
| 2-Inch Shower Drain | 1.60 | 1.60 | 3.20 |
| 4-Inch Cast Iron Arca Drain | 1.60 | 1.60 | 3.20 |
| 4-Inch Cast Iron Floor Drain with Trap | 3.20 | 3.20 | 6.40 |
| 4-Inch Roof Drain with Expansion Joint | 4.80 | 4.80 | 9.60 |
| 6-Person Wash Fountain, 54-Inch Semicircle | 8.00 | 8.00 | 16.00 |
| 4-Person Wash Fountain, 36-Inch Semicircle | 6.40 | 6.40 | 12.80 |
| 6-Person Wash Fountain, 36-Inch Round | 6.40 | 6.40 | 12.80 |
| 10-Person Wash Fountain, 54-Inch Round | 8.00 | 8.00 | 16.00 |
| Drinking Fountain | 3.20 | 3.20 | 6.40 |
| Eyewash | 1.60 | 1.60 | 3.20 |

Manhours include handling, hauling and installing fixtures, fixture fittings and drains as outlined above.

Manhours do not include earthwork or roughing-in. See respective tables for these charges.

# INSTALL BATHROOM ACCESSORIES

NET MANHOURS EACH

| Item | MANHOURS |
|---|---|
| | Carpenter |
| Mirrors | |
| Small | .25 |
| Medium | .75 |
| Large | 1.50 |
| Medicine Cabinets | .65 |
| Wall Towel Dispenser | .65 |
| Towel Bar | .05 |
| Wall Waste Receptacle | .65 |
| Sanitary Napkin Dispenser | .65 |
| Sanitary Napkin Disposal | .65 |
| Soap Dish | .05 |
| Paper Holder | .05 |

Manhours include handling, hauling, and complete installation of items as outlined above.

# TOILET AND SHOWER PARTITIONS

NET MANHOURS EACH

| Item | Manhours |
|---|---|
| | Carpenter |
| Pre Fabricated Toiled Partitions | |
| Walls incl. bracket & head rail | 2.18 |
| Stiles or pilaster | 0.40 |
| Posts | 0.76 |
| Doors – No Paylock | 0.40 |
| Toilet or Shower Stall Partitions | |
| Toilet stall with door | 1.94 |
| Toilet stall without door | 1.64 |
| Shower Stall | 1.23 |
| Urinal screen | 0.98 |

Manhours include handling, hauling and complete installation of items as outlined above.

# ROOF AND FLOOR DRAINS

NET MANHOURS EACH

| Item | MANHOURS | | |
|---|---|---|---|
| | Plumber | Helper | Total |
| <u>Roof Drains</u> | | | |
| Dome type | 1.75 | 1.75 | 3.50 |
| Flat type with strainer | 1.75 | 1.75 | 3.50 |
| <u>Cast Iron Floor Drains</u> | | | |
| 2" with strainer | 0.80 | 0.80 | 1.60 |
| 3" with strainer | 0.96 | 0.96 | 1.92 |
| 4" with sediment bucket | 1.28 | 1.28 | 2.56 |

**Manhours include handling, hauling and installing drains as outlined above.**

**Manhours do not include earthwork or roughing-in of pipe. See respective tables for these charges.**

# GALVANIZED SHEET METAL DRAINAGE

MANHOURS PER UNITS LISTED

| Item | Unit | MANHOURS | | |
|---|---|---|---|---|
| | | Sheet Metal Worker | Helper | Total |
| Gutter or Eve Troughs | | | | |
| Fabricate | lin. ft. | .08 | .08 | .16 |
| Erect | lin. ft. | .06 | .06 | .12 |
| Conductor Pipes or Downspouts | | | | |
| Fabricate | lin. ft. | .08 | .08 | .16 |
| Erect | lin. ft. | .05 | .05 | .10 |
| Elbows and Shoes | | | | |
| Fabricate | each | .08 | .03 | .09 |
| Erect | each | .04 | .01 | .05 |
| Erect Wire Conductor Strainers | each | .05 | .02 | .07 |

Manhours include handling, hauling, fabricating or erecting items outlined above.

Manhours do not include earthwork or installing underground drainage. See respective tables for these charges.

# COPPER DRAINAGE

MANHOURS PER UNITS LISTED

| Item | Unit | MANHOURS | | |
|---|---|---|---|---|
| | | Sheet Metal Worker | Helper | Total |
| Gutter or Eve Troughs | | | | |
| Fabricate | lin. ft. | .09 | .09 | .18 |
| Erect | lin. ft. | .07 | .07 | .14 |
| Conductor Pipes or downspouts | | | | |
| Fabricate | lin. ft. | .09 | .09 | .18 |
| Erect | lin. ft. | .06 | .06 | .12 |
| Elbows and Shoes | | | | |
| Fabricate | each | .10 | .04 | .14 |
| Erect | each | .05 | .01 | .06 |
| Erect Copper Conductor Strainers | each | .06 | .02 | .08 |

Manhours include handling, hauling and fabricating or erecting of items as outlined above.

Manhours do not include earthwork or installation of underground drainage. See respective tables for these charges.

# Section 2

# DUCTWORK

This section covers the manhours required for fabricating and erecting ducts and duct accessories for a heating, ventilating, or air-conditioning system in an industrial or chemical plant.

Ductwork is normally estimated on a weight basis for handling, hauling, fabricating, and erecting. By converting this to square or linear feet of ductwork, fittings by each, etc., a more realistic time frame for the various labor operations can be obtained. Therefore, the manhour tables in this section have been listed in this form. These can be converted to a per pound basis if desired.

The manhours listed are for labor only and have no bearing on material cost. All labor for handling, hauling, rigging, hoisting, setting, and aligning has been given due consideration in the manhour tables.

# SHOP FABRICATION – DUCTS

## Rectangular Sheet Metal Steel

MANHOURS PER SQUARE FOOT OF DUCT SURFACE

| Air Supply Cu. Ft. Per Minute | Perimeter Feet | Metal U. S. Gauge | Manhours Per Sq. Ft. of Duct Surface |
|---|---|---|---|
| 1000 | 4.0 | 24 | .12 |
| 3000 | 6.4 | 24 | .12 |
| 5000 | 8.0 | 24 | .13 |
| 7000 | 10.0 | 22 | .13 |
| 9000 | 11.2 | 22 | .13 |
| 10000 | 12.0 | 20 | .14 |
| 15000 | 14.6 | 20 | .16 |
| 20000 | 16.0 | 20 | .17 |
| 25000 | 17.0 | 18 | .18 |
| 50000 | 20.0 | 18 | .19 |
| 75000 | 25.0 | 18 | .19 |
| 100000 | 28.6 | 18 | .19 |
| 125000 | 32.0 | 18 | .20 |
| 150000 | 34.5 | 18 | .20 |

**Manhours include handling, hauling to fabricating shop, fabricating and hauling to storage yard or erection site ready for erection.**

**Manhours are for fabrication of straight sections of rectangular steel ducts only and do not include time for the fabrication of duct fittings or accessories or the erection of ducts, fittings or accessories. See respective tables for these charges.**

***Note:*** **Air supply column is based on a low-pressure system with air-friction loss being 0.08-inch per 100 lineal feet.**

# SHOP FABRICATION – DUCTS

## Rectangular Aluminum Sheet Metal

MANHOURS PER SQUARE FOOT, OF DUCT SURFACE

| Air Supply Cu. Ft. per Minutes | Perimeter Feet | Metal B & S Gauge | Manhours Per Sq. Ft. of Duct Surface |
|---|---|---|---|
| 1000 | 4.0 | 22 | .14 |
| 3000 | 6.4 | 22 | .14 |
| 5000 | 8.0 | 22 | .16 |
| 7000 | 10.0 | 20 | .16 |
| 9000 | 11.2 | 20 | .16 |
| 10000 | 12.0 | 18 | .17 |
| 15000 | 14.6 | 18 | .19 |
| 20000 | 16.0 | 18 | .20 |
| 25000 | 17.0 | 16 | .22 |
| 50000 | 20.0 | 16 | .23 |
| 75000 | 25.0 | 16 | .23 |
| 100000 | 28.6 | 16 | .23 |
| 125000 | 32.0 | 16 | .24 |
| 150000 | 34.5 | 16 | .24 |

Manhours include handling, hauling to fabricating shop, fabricating and hauling to storage yard or erection site ready for erection.

Manhours are for fabrication of straight sections of rectangular aluminum ducts only, and do not include time for the fabrication of duct fittings or accessories or the erection of ducts, fittings or accessories. See respective tables for these charges.

*Note:* Air supply column is based on air-friction loss of 0.08-inch water gauge per 100 linear feet.

# SHOP FABRICATION – DUCTS

## Rectangular Copper Sheet Metal

MANHOURS PER SQUARE FOOT OF DUCT SURFACE

| Air Supply Cu. Ft. Per Minute | Perimeter Feet | Metal B & S Gauge | Manhours Per Sq. Ft. of Duct Surface |
|---|---|---|---|
| 1000 | 4.0 | 22 | .14 |
| 3000 | 6.4 | 22 | .14 |
| 5000 | 8.0 | 22 | .15 |
| 7000 | 10.0 | 20 | .15 |
| 9000 | 11.2 | 20 | .15 |
| 10000 | 12.0 | 18 | .16 |
| 15000 | 14.6 | 18 | .18 |
| 20000 | 16.0 | 18 | .20 |
| 25000 | 17.0 | 16 | .21 |
| 50000 | 20.0 | 16 | .22 |
| 75000 | 25.0 | 16 | .22 |
| 100000 | 28.6 | 16 | .22 |
| 125000 | 32.0 | 16 | .23 |
| 150000 | 34.5 | 16 | .23 |

Manhours include handling, hauling to fabricating shop, fabricating and hauling to storage yard or erection site ready for erection.

Manhours are for fabrication of straight sections of rectangular sheet copper ducts only and do not include time for the fabrication of duct fittings or accessories or the erection of duct, fittings, or accessories. See respective tables for these charges.

*Note:* Air supply column is based on air-friction loss of 0.08-inch water gauge per 100 linear feet.

# SHOP FABRICATING AIR DUCTS

## Circular Primary Sheet Metal

MANHOURS PER LINEAR FOOT

| Duct Diameter Inches | Square Feet of Surface per Linear Foot | Sheet Gauge | Manhours Per Linear Foot |
|---|---|---|---|
| 4 | 1.05 | 22 | .07 |
| 6 | 1.57 | 22 | .11 |
| 8 | 2.10 | 22 | .15 |
| 10 | 2.62 | 22 | .19 |
| 12 | 3.14 | 20 | .25 |
| 14 | 3.65 | 20 | .29 |
| 16 | 4.20 | 20 | .34 |
| 18 | 4.65 | 20 | .38 |
| 20 | 5.25 | 20 | .42 |
| 21 | 5.50 | 18 | .49 |
| 24 | 6.30 | 18 | .61 |
| 26 | 6.80 | 18 | .67 |
| 28 | 7.30 | 18 | .71 |
| 30 | 7.85 | 18 | .77 |
| 31 | 8.13 | 16 | .94 |
| 36 | 9.45 | 16 | 1.10 |
| 40 | 10.50 | 16 | 1.22 |
| 41 | 10.70 | 14 | 1.52 |

Manhours include handling, hauling to fabricating shop, fabricating high-pressure air-conditioning ducts having longitudinal lock seams, cold cemented before locking and hauling to storage yard or erection site ready for erection.

Manhours are for fabrication of straight sections of duct only and do not include fabrication of duct fittings or erection of ducts or fittings. See respective tables for these charges.

# SHOP FABRICATING DUCT FITTINGS

## Rectangular Sheet Metal Steel

NET MANHOURS EACH

| Air Supply Cu. Ft. Per Minute | Per-imeter Feet | Metal U.S. Gauge | MANHOURS | | | | |
|---|---|---|---|---|---|---|---|
| | | | Ells | Tees | Wyes | Tap-ins | Offset Tran-sitions |
| 1000 | 4.0 | 24 | .16 | .24 | .28 | .16 | .20 |
| 3000 | 6.4 | 24 | .26 | .39 | .46 | .26 | .33 |
| 5000 | 8.0 | 24 | .32 | .48 | .56 | .32 | .40 |
| 7000 | 10.0 | 22 | .40 | .60 | .70 | .40 | .50 |
| 9000 | 11.2 | 22 | .45 | .68 | .79 | .45 | .56 |
| 10000 | 12.0 | 20 | .48 | .72 | .84 | .48 | .60 |
| 15000 | 14.6 | 20 | .58 | .87 | 1.02 | .58 | .73 |
| 20000 | 16.0 | 20 | .64 | .96 | 1.12 | .64 | .80 |
| 25000 | 17.0 | 18 | .68 | 1.02 | 1.19 | .68 | .85 |
| 50000 | 20.0 | 18 | .80 | 1.20 | 1.40 | .80 | 1.00 |
| 75000 | 25.0 | 18 | 1.00 | 1.50 | 1.75 | 1.00 | 1.25 |
| 100000 | 28.6 | 18 | 1.14 | 1.71 | 2.00 | 1.14 | 1.43 |
| 125000 | 32.0 | 18 | 1.28 | 1.92 | 2.24 | 1.28 | 1.60 |
| 150000 | 34.5 | 18 | 1.38 | 2.07 | 2.42 | 1.38 | 1.73 |

Manhours include handling, hauling, and shop fabrication of duct fittings as outlined above.

Manhours do not include fabrication of duct, duct accessories, or the installation of ducts, duct fittings, or duct accessories. See respective tables for these charges.

# SHOP FABRICATING DUCT FITTINGS

## Rectangular Sheet Metal Aluminum

NET MANHOURS EACH

| Air Supply Cu. Ft. Per Minute | Per-imeter Feet | Metal B & S Gauge | MANHOURS | | | | |
|---|---|---|---|---|---|---|---|
| | | | Ells | Tees | Wyes | Tap-ins | Offset Tran-sitions |
| 1000 | 4.0 | 22 | .20 | .30 | .35 | .20 | .25 |
| 3000 | 6.4 | 22 | .33 | .49 | .58 | .33 | .41 |
| 5000 | 8.0 | 22 | .40 | .60 | .70 | .40 | .50 |
| 7000 | 10.0 | 20 | .50 | .75 | .88 | .50 | .63 |
| 9000 | 11.2 | 20 | .56 | .85 | .99 | .56 | .70 |
| 10000 | 12.0 | 18 | .60 | .90 | 1.05 | .60 | .75 |
| 15000 | 14.6 | 18 | .73 | 1.09 | 1.28 | .73 | .91 |
| 20000 | 16.0 | 18 | .80 | 1.20 | 1.40 | .80 | 1.00 |
| 25000 | 17.0 | 16 | .85 | 1.28 | 1.49 | .85 | 1.06 |
| 50000 | 20.0 | 16 | 1.00 | 1.50 | 1.75 | 1.00 | 1.25 |
| 75000 | 25.0 | 16 | 1.25 | 1.87 | 2.19 | 1.25 | 1.56 |
| 100000 | 28.6 | 16 | 1.43 | 2.14 | 2.50 | 1.43 | 1.79 |
| 125000 | 32.0 | 16 | 1.60 | 2.40 | 2.80 | 1.60 | 2.00 |
| 150000 | 34.5 | 16 | 1.73 | 2.59 | 3.03 | 1.73 | 2.16 |

Manhours include handling, hauling, and shop fabrication of duct fittings as outlined above.

Manhours do not include fabrication of duct, duct accessories, or the installation of ducts, duct fittings, or duct accessories. See respective tables for these charges.

# SHOP FABRICATIONS DUCT FITTINGS

## Rectangular Sheet Metal Copper

NET MANHOURS EACH

| Air Supply Cu. Ft. Per Minute | Perimeter Feet | Metal B & S Gauge | MANHOURS | | | | |
|---|---|---|---|---|---|---|---|
| | | | Ells | Tees | Wyes | Tap-ins | Offset Transitions |
| 1000 | 4.0 | 22 | .18 | .28 | .32 | .18 | .23 |
| 3000 | 6.4 | 22 | .30 | .45 | .53 | .30 | .38 |
| 5000 | 8.0 | 22 | .37 | .55 | .64 | .37 | .46 |
| 7000 | 10.0 | 20 | .46 | .69 | .81 | .46 | .58 |
| 9000 | 11.2 | 20 | .52 | .78 | .91 | .52 | .64 |
| 10000 | 12.0 | 18 | .55 | .83 | .97 | .55 | .69 |
| 15000 | 14.6 | 18 | .67 | 1.00 | 1.17 | .67 | .84 |
| 20000 | 16.0 | 18 | .74 | 1.10 | 1.29 | .74 | .92 |
| 25000 | 17.0 | 16 | .78 | 1.17 | 1.37 | .78 | .98 |
| 50000 | 20.0 | 16 | .92 | 1.38 | 1.61 | .92 | 1.15 |
| 75000 | 25.0 | 16 | 1.15 | 1.73 | 2.01 | 1.15 | 1.44 |
| 100000 | 28.6 | 16 | 1.31 | 1.97 | 2.30 | 1.31 | 1.65 |
| 125000 | 32.0 | 16 | 1.47 | 2.21 | 2.58 | 1.47 | 1.84 |
| 150000 | 34.5 | 16 | 1.59 | 2.38 | 2.78 | 1.59 | 1.99 |

Manhours include handling, hauling, and shop fabrication of duct fittings as outlined above.

Manhours do not include fabrication of duct, duct accessories, or the installation of ducts, duct fittings, or duct accessories. See respective tables for these charges.

# SHOP FABRICATION AIR DUCT FITTINGS

## Circular Primary Sheet Metal

NET MANHOURS EACH

| Duct Diameter Inches | Square Feet of Surface Per Linear Foot | Sheet Gauge | MANHOURS | | |
|---|---|---|---|---|---|
| | | | Ells | Tees | Wyes |
| 4 | 1.05 | 22 | .20 | .27 | .45 |
| 6 | 1.57 | 22 | .20 | .27 | .45 |
| 8 | 2.10 | 22 | .20 | .27 | .45 |
| 10 | 2.62 | 22 | .20 | .27 | .45 |
| 12 | 3.14 | 20 | .20 | .27 | .45 |
| 14 | 3.65 | 20 | .25 | .33 | .55 |
| 16 | 4.20 | 20 | .25 | .33 | .55 |
| 18 | 4.65 | 20 | .28 | .37 | .61 |
| 20 | 5.25 | 20 | .28 | .37 | .61 |
| 21 | 5.50 | 18 | .30 | .40 | .66 |
| 24 | 6.30 | 18 | .30 | .40 | .66 |
| 26 | 6.80 | 18 | .30 | .40 | .66 |
| 28 | 7.30 | 18 | .35 | .47 | .78 |
| 30 | 7.85 | 18 | .35 | .47 | .78 |
| 31 | 8.13 | 16 | .38 | .51 | .85 |
| 36 | 9.45 | 16 | .38 | .51 | .85 |
| 40 | 10.50 | 16 | .40 | .53 | .88 |
| 41 | 10.71 | 14 | .40 | .53 | .88 |

Manhours include handling, hauling, and shop fabrication of duct fittings as outlined above.

Manhours do not include fabrication of duct, duct accessories, or the installation of ducts, duct fittings, or duct accessories. See respective tables for these charges.

# FIELD ERECTION DUCTS

## Rectangular Steel Sheet Metal

MANHOURS PER SQUARE FOOT OF DUCT SURFACE

| Air Supply Cu. Ft. per Minute | Perimeter Feet | Metal U. S. Gauge | Manhours Per Sq. Ft. of Duct surface |
|---|---|---|---|
| 1000 | 4.0 | 24 | .18 |
| 3000 | 6.4 | 24 | .18 |
| 5000 | 8.0 | 24 | .19 |
| 7000 | 10.0 | 22 | .19 |
| 9000 | 11.2 | 22 | .20 |
| 10000 | 12.0 | 20 | .22 |
| 15000 | 14.6 | 20 | .23 |
| 20000 | 16.0 | 20 | .25 |
| 25000 | 17.0 | 18 | .27 |
| 50000 | 20.0 | 18 | .28 |
| 75000 | 25.0 | 18 | .29 |
| 100000 | 28.6 | 18 | .29 |
| 125000 | 32.0 | 18 | .31 |
| 150000 | 34.5 | 18 | .31 |

Manhours include handling, rigging, hoisting into position, aligning and fastening into place.

Manhours do not include fabrication or installation of fittings or accessories. See respective tables for these charges.

# FIELD ERECTION DUCTS

## Rectangular Aluminum Sheet Metal

MANHOURS PER SQUARE FOOT OF DUCT SURFACE

| Air Supply Cu. Ft. Per Minute | Perimeter Feet | Metal B & S Gauge | Manhours Per Sq. Ft. of Duct Surface |
|---|---|---|---|
| 1000 | 4.0 | 22 | .22 |
| 3000 | 6.4 | 22 | .22 |
| 5000 | 8.0 | 22 | .23 |
| 7000 | 10.0 | 20 | .23 |
| 9000 | 11.2 | 20 | .24 |
| 10000 | 12.0 | 18 | .26 |
| 15000 | 14.6 | 18 | .28 |
| 20000 | 16.0 | 18 | .30 |
| 25000 | 17.0 | 16 | .32 |
| 50000 | 20.0 | 16 | .34 |
| 75000 | 25.0 | 16 | .35 |
| 100000 | 28.6 | 16 | .35 |
| 125000 | 32.0 | 16 | .37 |
| 150000 | 34.5 | 16 | .37 |

Manhours include handling, rigging, hoisting into position aligning and fastening into place.

Manhours do not include fabrication or the installation of fittings or accessories. See respective tables for these charges.

# FIELD ERECTION DUCTS

## Rectangular Copper Sheet Metal

MANHOURS PER SQUARE FOOT OF DUCT SURFACE

| Air Supply Cu. Ft. per Minute | Perimeter Feet | Metal B & S Gauge | Manhours per Sq. Ft. of Duct Surface |
|---|---|---|---|
| 1000 | 4.0 | 22 | .21 |
| 3000 | 6.4 | 22 | .21 |
| 5000 | 8.0 | 22 | .22 |
| 7000 | 10.0 | 20 | .22 |
| 9000 | 11.2 | 20 | .23 |
| 10000 | 12.0 | 18 | .25 |
| 15000 | 14.6 | 18 | .26 |
| 20000 | 16.0 | 18 | .29 |
| 25000 | 17.0 | 16 | .31 |
| 50000 | 20.0 | 16 | .32 |
| 75000 | 25.0 | 16 | .33 |
| 100000 | 28.6 | 16 | .33 |
| 125000 | 32.0 | 16 | .36 |
| 150000 | 34.5 | 16 | .36 |

Manhours include handling, rigging, hoisting into position, aligning and fastening into place.

Manhours do not include fabrication, or installation of fittings or accessories. See respective tables for these charges.

# FIELD ERECTION AIR DUCTS

## Circular Primary Sheet Metal

MANHOURS PER LINEAR FOOT

| Duct Diameter Inches | Square Feet of Surface Per Linear Foot | Sheet Gauge | Manhours Per Linear Foot |
|---|---|---|---|
| 4 | 1.05 | 22 | .11 |
| 6 | 1.57 | 22 | .17 |
| 8 | 2.10 | 22 | .23 |
| 10 | 2.62 | 22 | .28 |
| 12 | 3.14 | 20 | .38 |
| 14 | 3.65 | 20 | .44 |
| 16 | 4.20 | 20 | .50 |
| 18 | 4.65 | 20 | .56 |
| 20 | 5.25 | 20 | .63 |
| 21 | 5.50 | 18 | .73 |
| 24 | 6.30 | 18 | .91 |
| 26 | 6.80 | 18 | 1.00 |
| 28 | 7.30 | 18 | 1.07 |
| 30 | 7.85 | 18 | 1.15 |
| 31 | 8.13 | 16 | 1.41 |
| 60 | 9.45 | 16 | 1.64 |
| 40 | 10.50 | 16 | 1.83 |
| 41 | 10.70 | 14 | 2.28 |

Manhours include handling, rigging, hoisting into position, aligning and fastening into place.

Manhours do not include fabrication or installation of fittings or accessories. See respective tables for these charges.

# FIELD ERECTION DUCT FITTINGS

## Rectangular Sheet Metal Steel

NET MANHOURS EACH

| Air Supply Cu. Ft. Per Minute | Per-imeter Feet | Metal U.S. Gauge | MANHOURS | | | | |
|---|---|---|---|---|---|---|---|
| | | | Ells | Tees | Wyes | Tap-ins | Offset Tran-sitions |
| 1000 | 4.0 | 24 | .24 | .36 | .42 | .24 | .30 |
| 3000 | 6.4 | 24 | .39 | .59 | .69 | .39 | .50 |
| 5000 | 8.0 | 24 | .48 | .72 | .84 | .48 | .60 |
| 7000 | 10.0 | 22 | .60 | .90 | 1.05 | .60 | .75 |
| 9000 | 11.2 | 22 | .68 | 1.02 | 1.19 | .68 | .84 |
| 10000 | 12.0 | 20 | .72 | 1.08 | 1.26 | .72 | .90 |
| 15000 | 14.6 | 20 | .87 | 1.31 | 1.53 | .87 | 1.10 |
| 20000 | 16.0 | 20 | .96 | 1.44 | 1.68 | .96 | 1.20 |
| 25000 | 17.0 | 18 | 1.02 | 1.53 | 1.79 | 1.02 | 1.28 |
| 50000 | 20.0 | 18 | 1.20 | 1.80 | 2.10 | 1.20 | 1.50 |
| 75000 | 25.0 | 18 | 1.50 | 2.25 | 2.63 | 1.50 | 1.88 |
| 100000 | 28.6 | 18 | 1.71 | 2.57 | 3.00 | 1.71 | 2.15 |
| 125000 | 32.0 | 18 | 1.92 | 2.88 | 3.36 | 1.92 | 2.40 |
| 150000 | 34.5 | 18 | 2.07 | 3.11 | 3.63 | 2.07 | 2.60 |

Manhours include handling, hoisting into place, setting, aligning into position, and make-up of joints for the type fittings as listed above.

Manhours do not include fabrication or the placement of duct or accessories. See respective tables for these charges.

# FIELD ERECTION DUCT FITTINGS

## Rectangular Sheet Metal Aluminum

NET MANHOURS EACH

| Air Supply Cu. Ft. Per Minute | Per-imeter Feet | Metal B & S Gauge | MANHOURS | | | | |
|---|---|---|---|---|---|---|---|
| | | | Ells | Tees | Wyes | Tap-ins | Offset Tran-sitions |
| 1000 | 4.0 | 22 | .30 | .45 | .53 | .30 | .38 |
| 3000 | 6.4 | 22 | .50 | .74 | .87 | .50 | .62 |
| 5000 | 8.0 | 22 | .60 | .90 | 1.05 | .60 | .75 |
| 7000 | 10.0 | 20 | .75 | 1.13 | 1.32 | .75 | .95 |
| 9000 | 11.2 | 20 | .84 | 1.28 | 1.49 | .84 | 1.05 |
| 10000 | 12.0 | 18 | .90 | 1.35 | 1.58 | .90 | 1.13 |
| 15000 | 14.6 | 18 | 1.10 | 1.64 | 1.92 | 1.10 | 1.37 |
| 20000 | 16.0 | 18 | 1.20 | 1.80 | 2.10 | 1.20 | 1.50 |
| 25000 | 17.0 | 16 | 1.28 | 1.92 | 2.24 | 1.28 | 1.59 |
| 50000 | 20.0 | 16 | 1.50 | 2.25 | 2.63 | 1.50 | 1.88 |
| 75000 | 25.0 | 16 | 1.88 | 2.81 | 3.29 | 1.88 | 2.34 |
| 100000 | 28.6 | 16 | 2.15 | 3.21 | 3.75 | 2.15 | 2.69 |
| 125000 | 32.0 | 16 | 2.40 | 3.60 | 4.20 | 2.40 | 3.00 |
| 150000 | 34.5 | 16 | 2.60 | 3.89 | 4.55 | 2.60 | 3.24 |

Manhours include handling, hoisting into place, setting, aligning into position, and make-up of joints for the type fittings as listed above.

Manhours do not include fabrication or the placement of duct or accessories. See respective tables for these charges.

# FIELD ERECTION DUCT FITTINGS

## Rectangular Sheet Metal Copper

NET MANHOURS EACH

| Air Supply Cu. Ft. Per Minute | Per-imeter Feet | Metal B & S Gauge | MANHOURS | | | | |
|---|---|---|---|---|---|---|---|
| | | | Ells | Tees | Wyes | Tap-ins | Offset Tran-sitions |
| 1000 | 4.0 | 22 | .27 | .42 | .48 | .27 | .35 |
| 3000 | 6.4 | 22 | .45 | .68 | .80 | .45 | .57 |
| 5000 | 8.0 | 22 | .56 | .83 | .96 | .56 | .69 |
| 7000 | 10.0 | 20 | .69 | 1.04 | 1.22 | .69 | .87 |
| 9000 | 11.2 | 20 | .78 | 1.17 | 1.37 | .78 | .96 |
| 10000 | 12.0 | 18 | .83 | 1.25 | 1.46 | .83 | 1.04 |
| 15000 | 14.6 | 18 | 1.00 | 1.50 | 1.76 | 1.00 | 1.26 |
| 20000 | 16.0 | 18 | 1.11 | 1.65 | 1.94 | 1.11 | 1.38 |
| 25000 | 17.0 | 16 | 1.17 | 1.76 | 2.06 | 1.17 | 1.47 |
| 50000 | 20.0 | 16 | 1.38 | 2.07 | 2.42 | 1.38 | 1.73 |
| 75000 | 25.0 | 16 | 1.73 | 2.60 | 3.02 | 1.73 | 2.16 |
| 100000 | 28.6 | 16 | 1.97 | 2.96 | 3.45 | 1.97 | 2.48 |
| 125000 | 32.0 | 16 | 2.21 | 3.32 | 3.87 | 2.21 | 2.76 |
| 150000 | 34.5 | 16 | 2.39 | 3.57 | 4.17 | 2.39 | 2.99 |

Manhours include handling, hoisting into place, setting, aligning into position, and make-up of joints for the type fittings as listed above.

Manhours do not include fabrication or the placement of duct or accessories. See respective tables for these charges.

# FIELD ERECTION AIR DUCT FITTINGS

## Rectangular Aluminum Sheet Metal

NET MANHOURS EACH

| Duct Diameter Inches | Square Feet of Surface Per Linear Foot | Sheet Gauge | MANHOURS | | |
|---|---|---|---|---|---|
| | | | Ells | Tees | Wyes |
| 4 | 1.05 | 22 | .30 | .41 | .68 |
| 6 | 1.57 | 22 | .30 | .41 | .68 |
| 8 | 2.10 | 22 | .30 | .41 | .68 |
| 10 | 2.62 | 22 | .30 | .41 | .68 |
| 12 | 3.14 | 20 | .30 | .41 | .68 |
| 14 | 3.65 | 20 | .38 | .50 | .83 |
| 16 | 4.20 | 20 | .38 | .50 | .83 |
| 18 | 4.65 | 20 | .42 | .56 | .92 |
| 20 | 5.25 | 20 | .42 | .56 | .92 |
| 21 | 5.50 | 18 | .45 | .60 | .99 |
| 24 | 6.30 | 18 | .45 | .60 | .99 |
| 26 | 6.80 | 18 | .45 | .60 | .99 |
| 28 | 7.30 | 18 | .53 | .71 | 1.17 |
| 30 | 7.85 | 18 | .53 | .71 | 1.17 |
| 31 | 8.13 | 16 | .57 | .77 | 1.28 |
| 36 | 9.45 | 16 | .57 | .77 | 1.28 |
| 40 | 10.50 | 16 | .60 | .80 | 1.32 |
| 41 | 10.71 | 14 | .60 | .80 | 1.32 |

Manhours include handling, hoisting into place, setting, aligning into position, and make-up of joints for the type fittings as listed above.

Manhours do not include fabrication or the placement of duct or accessories. See respective tables for these charges.

# FIELD INSTALLATION DUCT ACCESSORIES

## Rectangular Sheet Metal Steel

NET MANHOURS EACH

| Air Supply Cu. Ft. per Minute | Perimeter Feet | Metal U. S. Gauge | MANHOURS | | | | |
|---|---|---|---|---|---|---|---|
| | | | Duct Enclosures For | | Duct Access Door | Split-ters | Volume Dampers |
| | | | Fin Coils | Sound Absorber | | | |
| 1000 | 4.0 | 24 | .07 | .06 | .10 | .06 | .10 |
| 3000 | 6.4 | 24 | .07 | .06 | .10 | .06 | .10 |
| 5000 | 8.0 | 24 | .09 | .08 | .12 | .08 | .12 |
| 7000 | 10.0 | 22 | .09 | .08 | .12 | .08 | .12 |
| 9000 | 11.2 | 22 | .09 | .08 | .12 | .08 | .12 |
| 10000 | 12.0 | 20 | .10 | .09 | .13 | .09 | .13 |
| 15000 | 14.6 | 20 | .10 | .09 | .13 | .09 | .13 |
| 20000 | 16.0 | 20 | .12 | .11 | .15 | .11 | .15 |
| 25000 | 17.0 | 18 | .12 | .11 | .15 | .11 | .15 |
| 50000 | 20.0 | 18 | .13 | .12 | .16 | .12 | .16 |
| 75000 | 25.0 | 18 | .13 | .12 | .16 | .12 | .16 |
| 100000 | 28.6 | 18 | .15 | .14 | .18 | .14 | .18 |
| 125000 | 32.0 | 18 | .15 | .14 | .18 | .14 | .18 |
| 150000 | 34.5 | 18 | .15 | .14 | .18 | .14 | .18 |

Manhours include handling, hoisting into place, setting, aligning into position and make-up of connections for the above type accessories.

Manhours do not include fabrication or the placement of duct or fittings. See respective tables for these charges.

# FIELD INSTALLATION DUCT ACCESSORIES

## Rectangular Sheet Metal Aluminum

NET MANHOURS EACH

| Air Supply Cu. Ft. Per Minute | Perimeter Feet | Metal B & S Gauge | MANHOURS | | | | |
|---|---|---|---|---|---|---|---|
| | | | Duct Enclosures For | | Duct Access Doors | Splitters | Volume Dampers |
| | | | Fin Coils | Sound Absorber | | | |
| 1000 | 4.0 | 22 | .08 | .07 | .12 | .07 | .12 |
| 3000 | 6.4 | 22 | .08 | .07 | .12 | .07 | .12 |
| 5000 | 8.0 | 22 | .11 | .10 | .14 | .10 | .14 |
| 7000 | 10.0 | 20 | .11 | .10 | .14 | .10 | .14 |
| 9000 | 11.2 | 20 | .11 | .10 | .14 | .10 | .14 |
| 10000 | 12.0 | 18 | .12 | .11 | .16 | .11 | .16 |
| 15000 | 14.6 | 18 | .12 | .11 | .16 | .11 | .16 |
| 20000 | 16.0 | 18 | .14 | .13 | .18 | .13 | .18 |
| 25000 | 17.0 | 16 | .14 | .13 | .18 | .13 | .18 |
| 50000 | 20.0 | 16 | .16 | .14 | .19 | .14 | .19 |
| 75000 | 25.0 | 16 | .16 | .14 | .19 | .14 | .19 |
| 100000 | 28.6 | 16 | .18 | .17 | .22 | .17 | .22 |
| 125000 | 32.0 | 16 | .18 | .17 | .22 | .17 | .22 |
| 150000 | 34.5 | 16 | .18 | .17 | .22 | .17 | .22 |

Manhours include handling, hoisting into place, setting, aligning into position and make-up of connections for the above type accessories.

Manhours do not include fabrication or the placement of duct or fittings. See respective tables for these charges.

# FIELD INSTALLATION DUCT ACCESSORIES

## Rectangular Sheet Metal Copper

NET MANHOURS EACH

| Air Supply Cu. Ft. Per Minute | Perimeter Feet | Metal B & S Gauge | MANHOURS | | | | |
|---|---|---|---|---|---|---|---|
| | | | DUCT ENCLOSURES | | Duct Access Door | Splitters | Volume Dampers |
| | | | Fin Coils | Sound Absorber | | | |
| 1000 | 4.0 | 22 | .09 | .08 | .14 | .08 | .14 |
| 3000 | 6.4 | 22 | .09 | .08 | .14 | .08 | .14 |
| 5000 | 8.0 | 22 | .13 | .12 | .16 | .12 | .16 |
| 7000 | 10.0 | 20 | .13 | .12 | .16 | .12 | .16 |
| 9000 | 11.2 | 20 | .13 | .12 | .16 | .12 | .16 |
| 10000 | 12.0 | 18 | .14 | .13 | .18 | .13 | .18 |
| 15000 | 14.6 | 18 | .14 | .13 | .18 | .13 | .18 |
| 20000 | 16.0 | 18 | .16 | .15 | .21 | .15 | .21 |
| 25000 | 17.0 | 16 | .16 | .15 | .21 | .15 | .21 |
| 50000 | 20.0 | 16 | .18 | .16 | .22 | .16 | .22 |
| 75000 | 25.0 | 16 | .18 | .16 | .22 | .16 | .22 |
| 100000 | 28.6 | 16 | .21 | .20 | .25 | .20 | .25 |
| 125000 | 32.0 | 16 | .21 | .20 | .25 | .20 | .25 |
| 150000 | 34.5 | 16 | .21 | .20 | .25 | .20 | .25 |

Manhours include handling, hoisting into place, setting, aligning into position and make-up of connections for the above type accessories.

Manhours do not include fabrication or the placement of duct or fittings. See respective tables for these charges.

# INSTALLATION OF STEEL CEILING DIFFUSERS & ACCESSORIES

## Ceiling Diffusers

MANHOURS REQUIRED EACH

| Size Inches | ROUND SHAPES | | | | SQUARE SHAPES | | |
|---|---|---|---|---|---|---|---|
| | Nominal CFM (43 DB) | Flush Mount | Step-Down Adjustable | Half-Round | Nominal CFM (43 DB) | Flush Mount | With Equalizing Grid & Damper |
| 6 | 225 | .60 | 1.20 | .40 | 250 | .60 | .74 |
| 8 | 400 | .60 | 1.20 | .40 | 400 | .60 | .74 |
| 10 | 600 | .65 | 1.25 | .44 | 600 | .65 | .79 |
| 12 | 750 | .70 | 1.30 | .47 | 750 | .70 | .84 |
| 14 | 1100 | .75 | 1.35 | .50 | – | .75 | .89 |
| 15 | – | – | – | – | 1100 | .80 | .94 |
| 16 | 1250 | .85 | 1.45 | .57 | – | .85 | .99 |
| 18 | 1550 | .90 | 1.50 | .60 | 1550 | .90 | 1.04 |
| 20 | 2100 | – | 1.60 | .67 | – | – | – |
| 24 | 2500 | – | 1.75 | .77 | – | – | – |
| 30 | 3400 | – | 1.90 | – | – | – | – |
| 38 | 5000 | – | 2.10 | – | – | – | – |

# ACCESSORIES FOR CEILING DIFFUSERS

MANHOURS REQUIRED EACH

| Size Inches | Equalizing Grid | Volume Extractor | Volume Damper | Smudge Ring |
|---|---|---|---|---|
| 6 | .14 | .12 | .14 | .12 |
| 8 | .14 | .12 | .14 | .12 |
| 10 | .16 | .14 | .16 | .13 |
| 12 | .16 | .14 | .16 | .13 |
| 14 | .16 | .14 | .16 | .14 |
| 15 | .18 | .16 | .18 | .14 |
| 16 | .18 | .16 | .18 | .16 |
| 18 | .21 | .18 | .21 | .16 |
| 20 | .21 | .20 | .21 | .18 |
| 24 | .22 | .20 | .22 | .18 |
| 30 | .22 | – | .22 | – |
| 38 | .25 | – | .25 | – |

Manhours include job handling, hauling, and installing of diffusers and accessories as outlined.

Manhours do not include scaffolding. See respective table for this time frame.

# FIELD INSTALLATION OF MISCELLANEOUS GRILLS

MANHOURS REQUIRED EACH

| Item and Sizes | Manhours Each |
|---|---|
| Ceiling Grills | |
| 6″ x 4″ through 10″ x 10″ | 0.70 |
| 12″ x 4″ through 12″ x 12″ | 0.80 |
| 14″ x 6″ through 14″ x 14″ | 0.90 |
| Side-Wall Supply Grills | |
| 8″ x 4″ through 12″ x 8″ | 0.35 |
| 14″ x 4″ through 18″ x 6″ | 0.40 |
| 20″ x 5″ through 24″ x 12″ | 0.48 |
| 30″ x 6″ through 36″ x 12″ | 0.55 |
| 40″ x 8″ through 48″ x 8″ | 0.60 |
| Return Air Grills | |
| 11″ x 6″ through 18″ x 18″ | 0.40 |
| 24″ x 10″ through 30″ x 24″ | 0.50 |
| 36″ x 12″ through 36″ x 30″ | 0.55 |
| 48″ x 24″ through 48″ x 36″ | 0.60 |

Manhours include job handling, hauling, and installing aluminum or steel grills.

*Ceiling Grill* manhours are for installation of rectangular-shaped grills with adjustable curved blades and are for one-way or two-way throw grills. For one-way throw grills with multi-shutter, increase manhours 10%.

*Side Wall Supply Grill* manhours are for installation of single deflection with multi-shutter damper control, double deflection with no damper control, and double deflection with opposed blades and damper control.

*Return Air Grill* manhours are for installation of standard wall grills with single facebars. For hinged grills with filter back, increase manhours 10%. For standard door, site-tite with frame grills increase manhours 25%. For installation of remote wall controller, including OB damper behind grill add 1.5 manhours each.

Manhours do not include hole cutting or scaffolding. See respective tables for these time frames.

# FIELD ERECT SKYLIGHTS AND ROOF VENTILATORS

MANHOURS PER UNITS LISTED

| Item | Unit | Manhours |
|---|---|---|
| Pre fabricated Metal Skylights: | | |
| Small – up to 100 square feet | each | 7.20 |
| Medium – 100 to 150 square feet | each | 12.00 |
| Large – 150 to 200 square feet | each | 18.00 |
| Gravity Roof ventilators | cwt. | 7.00 |
| Power Roof ventilators | cwt. | 7.20 |
| Ridge ventilators | lin. ft. | 0.20 |
| Louvers | | |
| Fabricate | sq. ft. | 0.12 |
| Install | sq. ft. | 0.15 |

Manhours include all labor necessary for handling, hauling and installing items listed above.

Prefabricated metal skylight units include allowance for glazing.

Manhours do not include scaffolding. See respective table for this charge.

# Section 3

# EQUIPMENT INSTALLATION

The following manhour tables cover the installation of equipment and other miscellaneous items for a heating, ventilating, or air-conditioning system in an industrial or chemical plant.

These tables include the manhours required to unload, handle, rig, hoist, set, and align various types of equipment, as well as the manhours to weld plate steel for tank fabrication.

Ths listed manhours include time allowance to complete all labor for the particular operation as outlined in the various tables and in accordance with the notes thereon.

# FIELD INSTALLATION

## Radiators and Convectors

MANHOURS PER HUNDRED (100) Square Feet

| Item | MANHOURS | | |
|---|---|---|---|
| | Pipefitter | Helper | Total |
| Floor Installed | | | |
| One pipe system | 1.84 | 1.84 | 3.68 |
| Two pipe system | 2.24 | 1.84 | 4.08 |
| Wall Installed | | | |
| One pipe system | 2.30 | 2.30 | 4.60 |
| Two pipe system | 2.80 | 2.80 | 5.60 |

**Manhours include handling, hauling and installing of items.**

**Manhours do not include service pipe installation. See respective tables for these charges.**

# TRUSS OR SUSPENDED FANS & MOTORS

NET MANHOURS PER UNIT

| Air Cu. Ft. Per Minute | Approximate Horsepower | Fan & Motor Weight in Pounds | Manhours |
|---|---|---|---|
| 1000-2000 | 1 or less | 400 | 6.0 |
| 3000 | 1-1/2 | 500 | 7.5 |
| 4000 | 2 | 600 | 9.0 |
| 6000 | 3 | 700 | 10.5 |
| 10000 | 5 | 1300 | 15.6 |
| 15000 | 7-1/2 | 1800 | 21.6 |
| 22000 | 10 | 2500 | 30.0 |
| 33000 | 15 | 3900 | 42.0 |
| 40000 | 20 | 5000 | 52.5 |
| 50000 | 25 | 6000 | 63.0 |

Manhours include handling, hauling, rigging, setting and aligning of fans and motors to 20 feet high as outlined above. Increase manhours by one half of one per cent for each foot above 20 feet.

Manhours do not include scaffolding or the installation of supports. See respective tables for these charges.

For sizes not listed use the next higher listing.

# TRUSS OR SUSPENDED HEATING & VENTILATING UNITS

NET MANHOURS PER UNIT

| 1000 BTU Per Hour Capacity | Approximate Air Cu. Ft. per Minute | Unit Weight in Pounds | Manhours |
|---|---|---|---|
| 50 | 700 | 300 | 8.0 |
| 120 | 1500 | 450 | 8.0 |
| 160 | 2000 | 500 | 8.0 |
| 180 | 3000 | 600 | 9.0 |
| 300 | 4000 | 900 | 10.8 |
| 400 | 4500 | 1500 | 18.0 |
| 500 | 5500 | 1600 | 19.2 |
| 750 | 8500 | 2500 | 30.0 |
| 1000 | 11000 | 2600 | 31.2 |
| 1250 | 14000 | 3500 | 42.0 |
| 1500 | 17000 | 3700 | 44.4 |
| 1750 | 19000 | 4300 | 51.6 |
| 2000 | 22000 | 4500 | 52.0 |

**Manhours include handling, hauling, rigging, setting and aligning heating and ventilating units to 20 feet high as outlined above. Increase manhours by one half of one per cent for each foot above 20 feet.**

**For sizes not listed use the next higher listing.**

**Manhours do not include scaffolding or the installation of supports. See respective tables for these charges.**

# TRUSS OR SUSPENDED CEILING SELF-CONTAINED AIR CONDITIONING UNITS

NET MANHOURS PER UNIT

| Refrigeration Tons | Air Cu. Ft. Per Minute | Air Conditioning Unit in Pounds | Manhours to Handle & Erect |
|---|---|---|---|
| 6 | 2000 | 2000 | 39.0 |
| 10 | 4000 | 3000 | 56.7 |
| 15 | 6000 | 3800 | 68.4 |
| 20 | 8000 | 4000 | 72.0 |
| 30 | 12000 | 4500 | 81.0 |
| 40 | 16000 | 6000 | 108.0 |
| 50 | 20000 | 8000 | 132.0 |

Manhours include handling, hauling, rigging, setting and aligning of self-contained air-conditioning units to 20 feet high as outlined above. Increase manhours by one half of one per cent for each foot above 20 feet.

Manhours do not include scaffolding or the installation of supports. See respective tables for these charges.

For sizes not listed use the next higher listing.

# AIR HANDLING UNITS— SINGLE ZONE & MULTI-ZONE

MANHOURS REQUIRED EACH

| Capacity CFM | Fan Horsepower | Conditioned Area Square Feet | Manhours |
|---|---|---|---|
| Single Zone Units | | | |
| 1,000 | 1.0 | 1,000 | 12.6 |
| 2,500 | 3.0 | 2,200 | 14.0 |
| 6,000 | 5.0 | 4,000 | 16.8 |
| 14,000 | 10.0 | 14,000 | 17.5 |
| 24,000 | 20.0 | 24,000 | 31.5 |
| 30,000 | 25.0 | 30,000 | 42.0 |
| Multi-Zone Units | | | |
| 4,000 | 3.0 | 4,000 | 12.6 |
| 6,000 | 5.0 | 6,000 | 15.4 |
| 10,000 | 7.5 | 10,000 | 16.8 |
| 15,000 | 15.0 | 15,000 | 25.2 |
| 22,000 | 20.0 | 22,000 | 28.0 |
| 30,000 | 25.0 | 30,000 | 42.0 |

The above units are based on a 550 FPM coil face velocity and 3 inches fan static pressure and include insulated casing, fan section, cooling coil section with a 6-row aluminum fin coil and drain pan, heating coil section with a 2-row aluminum fin coil, filter section with replaceable filters, fan motor, variable speed drive, and vibration isolators. Multi-zone units include zoning damper section.

Manhours include receiving at job site, off-loading from carrier, moving within 50 feet of final location site, uncrating, setting, and aligning.

Manhours do not include installation of ductwork, water or steam piping, motor starter or power wiring, and scaffolding. See respective tables for these time frames.

# ROOF-MOUNTED HEATING & COOLING UNITS

## COMBINATION UNITS

MANHOURS REQUIRED EACH

| Cooling Capacity Tons | Heating Capacity BTU/Hour | Power KW | Gas BTU/Hour | Nominal CFM | Average Weight Short Tons | Manhours |
|---|---|---|---|---|---|---|
| 5 | 112,500 | 8 | 150,000 | 2,000 | 0.50 | 18.0 |
| 7-1/2 | 165,000 | 12 | 225,000 | 3,000 | 0.60 | 19.2 |
| 10 | 206,000 | 16 | 275,000 | 4,000 | 1.10 | 20.1 |
| 15 | 270,000 | 19 | 360,000 | 6,000 | 1.15 | 21.0 |
| 20 | 360,000 | 25 | 480,000 | 8,000 | 1.80 | 24.0 |
| 30 | 540,000 | 40 | 720,000 | 12,000 | 2.65 | 25.6 |
| 40 | 675,000 | 50 | 900,000 | 16,000 | 3.20 | 26.8 |
| 50 | 835,000 | 62 | 1,115,000 | 20,000 | 3.65 | 28.0 |

# COMBINATION MULTI-ZONE MODULAR UNITS

MANHOURS REQUIRED EACH

| Cooling Capacity Tons | Quantity of Zone Modules | Heating Capacity BTU/Hour | Power KW | Gas BTU/Hour | Nominal CFM | Average Weight Short Tons | Manhours |
|---|---|---|---|---|---|---|---|
| 20 | 8 | 360,000 | 21 | 480,000 | 8,000 | 2.18 | 39.2 |
| 25 | 10 | 450,000 | 28 | 600,000 | 10,000 | 2.30 | 44.8 |
| 30 | 12 | 540,000 | 34 | 720,000 | 12,000 | 3.00 | 50.4 |

Manhours are based on receiving preassembled package type units, prewired, prepiped, and charged with refrigerant.

Units are electrically cooled, gas heated, and wired for 230 or 460 volts, 3-phase 60 Hz.

Installation of separately shipped roof curbs for units of 20 tons or greater is included.

Manhours do not include installation of power wiring or gas piping. See respective tables for these time frames.

# SPLIT SYSTEM AIR-COOLED PACKAGES

MANHOURS REQUIRED EACH

| Capacity Tons | Average Weight Short Tons | Manhours |
|---|---|---|
| 3 | 0.28 | 9.0 |
| 5 | 0.38 | 11.4 |
| 7-1/2 | 0.45 | 12.9 |
| 10 | 0.80 | 17.2 |
| 15 | 1.05 | 22.9 |
| 20 | 1.60 | 32.9 |
| 25 | 1.85 | 37.7 |
| 30 | 2.25 | 48.7 |
| 40 | 2.60 | 57.3 |
| 50 | 3.55 | 68.1 |

Manhours are based on receiving at job site, preassembled packages consisting of one high-side condensing unit and one low-side air handler companion piece, matched and ready for field piping.

Manhours include off-loading from carrier, moving within 50 feet of final location site, uncrating, setting, and aligning of the condensing unit on a preinstalled concrete slab along the exterior structure wall and suspending the air handler from the ceiling up to 15 feet high.

Manhours do not include installation of concrete slab, connecting piping, refrigerant charge, or scaffolding. See respective tables for these time frames.

# ROOM FAN COIL UNITS

MANHOURS REQUIRED EACH

| Capacity CFM | Cooling BTUH | Heating BTUH | Water GPM | Electric Watts | Weight Pounds | Manhours |
|---|---|---|---|---|---|---|
| 200 | 6,400 | 17,000 | 2 | 70 | 85 | 2.40 |
| 300 | 9,150 | 26,100 | 2 | 90 | 100 | 2.60 |
| 400 | 12,650 | 32,300 | 3 | 110 | 110 | 2.80 |
| 600 | 17,660 | 44,700 | 4 | 130 | 130 | 3.00 |

The above units include insulated cabinet with discharge air grill, 3-row aluminum fin coil, 115-volt, 1-phase motor, fan, 3-speed fan switch, thermostat, manual change-over switch, two-way solenoid valve, and throwaway filter.

Manhours include receiving at job site, off-loading from carrier, moving within 50 feet of final location, uncrating, setting, and aligning inside of building at grade.

If unit is to be supported from existing structural framing, up to 15 feet above floor, increase manhours 25%.

Manhours do not include installation of ductwork, water or steam piping, motor starter, or power wiring. See respective tables for these time frames.

# PACKAGED RECIPROCATING WATER CHILLERS WITH AIR-COOLED CONDENSERS

MANHOURS REQUIRED EACH

| Capacity Tons | Power KW | Average Weight Short Tons | Manhours |
|---|---|---|---|
| 20 | 23 | 1.35 | 27.7 |
| 30 | 35 | 1.60 | 28.4 |
| 40 | 53 | 2.10 | 29.5 |
| 50 | 62 | 2.95 | 32.8 |
| 60 | 90 | 3.05 | 34.2 |
| 80 | 97 | 4.70 | 37.8 |
| 100 | 116 | 5.10 | 38.6 |

# PACKAGED RECIPROCATING HERMETIC WATER CHILLERS WITH WATER-COOLED CONDENSERS

MANHOURS REQUIRED EACH

| Capacity Tons | Power KW | Average Weight Short Tons | Manhours |
|---|---|---|---|
| 15 | 14 | 0.80 | 21.4 |
| 30 | 29 | 1.00 | 21.9 |
| 50 | 48 | 1.75 | 22.6 |
| 60 | 59 | 1.80 | 32.8 |
| 80 | 76 | 2.70 | 33.5 |
| 100 | 93 | 3.25 | 37.8 |
| 120 | 112 | 3.35 | 38.6 |
| 150 | 154 | 4.00 | 39.9 |

Manhours are based on receiving package type units delivered to job site and include off-loading from carrier, moving within 50 feet of final location site, uncrating, setting, aligning, starting, and checking.

Manhours do not include installation of water supply and return piping, instruments within the water piping, or electrical power wiring. See respective tables for these time frames.

Air-cooled packages include direct expansion cooler, air-cooled condenser, condenser fan, starters for compressors and fan motors, operating and safety controls, insulation, internal wiring, refrigerant charge, and vibration eliminators.

Water-cooled packages include hermetic compressor, motor, cooler, condenser, internal piping and wiring, motor starters, insulation, operating and safety controls, refrigerant charge, and vibration eliminators.

# PACKAGED HERMETIC CENTRIFUGAL WATER CHILLERS WITH WATER-COOLED CONDENSERS

MANHOURS REQUIRED EACH

| Capacity Tons | Power KW | Average Weight Short Tons | Manhours |
|---|---|---|---|
| 100 | 95 | 3.80 | 37.8 |
| 130 | 115 | 3.95 | 38.6 |
| 150 | 130 | 4.90 | 39.5 |
| 200 | 165 | 5.75 | 40.3 |
| 250 | 215 | 6.40 | 42.8 |
| 300 | 250 | 7.75 | 43.8 |
| 350 | 285 | 8.05 | 44.7 |
| 400 | 345 | 9.00 | 45.7 |
| 500 | 395 | 10.00 | 47.9 |
| 600 | 465 | 11.00 | 48.9 |
| 800 | 620 | 15.00 | 52.9 |
| 1000 | 720 | 17.50 | 58.0 |

Manhours are based on receiving package type units delivered to job site and include off-loading from carrier, moving within 50 feet of final location site, uncrating, setting, aligning, starting, and checking.

Manhours do not include installation of water supply and return piping, instruments within the water piping, or electrical power wiring. See respective tables for these time frames.

Water-cooled packages include hermetic compressor, 460-volt motor, cooler, condenser, internal piping and wiring, purge units, gauges, controls, insulation, lubrication system, oil, and refrigerant charge.

# COOLING TOWERS FOR RECIPROCATING & CENTRIFUGAL CHILLERS

MANHOURS REQUIRED EACH

| Refrigeration Capacity Tons | Motor Horsepower | Average Weight Short Tons | Manhours |
|---|---|---|---|
| 20 | 3/4 | 1.00 | 40.0 |
| 30 | 1-1/2 | 1.11 | 43.0 |
| 40 | 1-1/2 | 1.20 | 55.0 |
| 50 | 2 | 1.30 | 61.0 |
| 65 | 3 | 1.50 | 76.0 |
| 85 | 3 | 2.00 | 96.0 |
| 100 | 5 | 2.25 | 122.0 |
| 200 | 10 | 2.90 | 244.0 |
| 300 | 15 | 4.00 | 330.0 |
| 400 | 20 | 5.00 | 413.0 |
| 600 | 25 | 6.50 | 558.0 |
| 800 | 30 | 8.00 | 698.0 |
| 1000 | 50 | 12.50 | 824.0 |

Manhours are based on receiving prefabricated units, knocked down, and delivered to job site and include off-loading from carrier, moving within 50 feet of final location site, field assembling, erecting, aligning, and anchoring at ground elevation.

If tower is to be set on structure roof, increase manhours 5%.

Manhours do not include installation of concrete basin, supporting steel, grillage, water treatment system, condenser water piping, motor starters, or power wiring. See respective tables for these time frames.

# WATER HEATING BOILERS—
# PACKAGE TYPE ELECTRIC HYDRONIC BOILERS

MANHOURS REQUIRED EACH

| Output BTU/Hour | Installation Manhours |
|---|---|
| 34,000 | 4.0 |
| 51,000 | 5.5 |
| 68,000 | 7.0 |
| 81,000 | 8.0 |

# CAST IRON GAS-FIRED BOILERS

MANHOURS REQUIRED EACH

| Net IBR Rating MBH | Installation Manhours |
|---|---|
| 46.1 | 13.2 |
| 68.7 | 17.6 |
| 104.3 | 22.0 |
| 156.5 | 27.4 |
| 208.7 | 36.5 |
| 260.9 | 41.0 |
| 313.0 | 45.6 |
| 365.8 | 47.9 |
| 417.4 | 52.8 |
| 469.6 | 57.6 |
| 556.5 | 70.2 |
| 695.7 | 75.4 |
| 834.8 | 93.0 |
| 973.9 | 96.0 |

Manhours include receiving at job site, off-loading from carrier, moving within 50 feet of final location, uncrating, setting, and aligning.

Manhours do not include installation of gas, steam, water, or other piping, or installation of motor starter, and power wiring. See respective tables for these time frames.

Electric hydronic boilers are complete packages including automatic temperature controls and are prewired for connection to 240-volt power source.

Cast iron gas-fired boilers include automatic gas value, fail-safe type pilot, manual shut-off, drain cock, low water cut-off, safety valve gauge glass, steam pressure gauge, and draft diverter.

# PACKAGE TYPE STEAM BOILERS

MANHOURS REQUIRED EACH

| Steam Output Pounds/Hour | Installation Manhours |
|---|---|
| 15 PSIG | |
| 1,000 | 88.0 |
| 1,500 | 96.8 |
| 2,000 | 110.4 |
| 2,500 | 119.6 |
| 3,750 | 134.4 |
| 5,000 | 144.0 |
| 150 PSIG | |
| 7,500 | 147.2 |
| 10,000 | 156.4 |
| 15,000 | 169.9 |
| 20,000 | 179.4 |

Manhours include off-loading from carrier at job site, skidding, rolling, jacking, setting, and aligning on pre-installed foundation, boilout, startup, final adjustment, and testing.

Manhours do not include setting of stack, electric power wiring to the unit, and gas, oil, steam, water, or other piping to and from the unit. See respective tables for these time requirements.

The above package type steam boilers are for a complete, fully factory-assembled, skid mounted, oil- or gas-fired, water tube boiler, and includes all trim and controls, condensate tank, boiler feed water pump, steam preheat system, and water softener for 100% make-up.

# ERECT PREFABRICATED BOILER STACKS

MANHOURS REQUIRED FOR UNITS LISTED

| Stack Diameter Inches | MANHOURS | |
|---|---|---|
| | Handle Stack Per Linear Feet | Flange-Up Each |
| 10 | 0.19 | 1.22 |
| 12 | 0.23 | 1.26 |
| 15 | 0.29 | 1.32 |
| 18 | 0.34 | 1.37 |
| 21 | 0.40 | 1.43 |
| 24 | 0.46 | 1.49 |
| 30 | 0.57 | 1.60 |
| 36 | 0.68 | 1.71 |

Manhours are based on job receiving of prefabricated, flanged, 15 to 20 feet long stack sections in a weight range of from 500 to 1000 pounds each.

Handle stack manhours include off-loading, rigging, picking, setting, aligning, and guying of stack sections.

Flange-up manhours include bolting of sections together.

# INSTALLATION OF HOT WATER HEATERS—AUTOMATIC GAS-FIRED

MANHOURS REQUIRED EACH

| Capacity Gallons | BTU | Weight Pounds | Installation Manhours |
|---|---|---|---|
| 20 | 22,000 | 109 | 3.00 |
| 30 | 33,000 | 135 | 3.25 |
| 40 | 36,000 | 160 | 3.50 |
| 50 | 50,000 | 242 | 4.25 |
| 60 | 62,000 | 275 | 5.00 |

# AUTOMATIC ELECTRIC

MANHOURS REQUIRED EACH

| Capacity Gallons | Upper Wattage | Lower Wattage | Weight Pounds | Installation Manhours |
|---|---|---|---|---|
| 30 | 1,250 | 750 | 121 | 2.25 |
| 40 | 1,500 | 1,000 | 145 | 2.44 |
| 50 | 1,500 | 1,000 | 201 | 3.15 |
| 65 | 2,000 | 1,000 | 290 | 3.40 |
| 80 | 2,500 | 1,500 | 309 | 4.00 |

Manhours include receiving at job site, off-loading from carrier, moving within 50 feet of final location, uncrating, setting, aligning, and hooking up.

Manhours do not include installation of gas or water piping or electric power. See respective tables for piping time frames. Electric power installation is assumed to be completed by electrical subcontractors.

# PREFABRICATED HOT WATER STORAGE TANKS

MANHOURS REQUIRED EACH

| Tank Size Inches | Capacity Gallons | Weight Pounds | Installation Manhours |
|---|---|---|---|
| 18 x 60 | 65 | 125 | 1.20 |
| 18 x 72 | 80 | 150 | 1.26 |
| 24 x 63 | 115 | 200 | 1.33 |
| 24 x 75 | 140 | 240 | 1.40 |
| 30 x 66 | 190 | 360 | 1.58 |
| 30 x 78 | 225 | 420 | 1.73 |
| 30 x 90 | 260 | 480 | 1.88 |
| 36 x 82 | 325 | 535 | 1.95 |
| 36 x 94 | 380 | 605 | 2.16 |
| 36 x 106 | 430 | 680 | 2.24 |
| 36 x 118 | 485 | 750 | 2.40 |
| 42 x 96 | 535 | 990 | 2.98 |
| 42 x 108 | 600 | 1100 | 3.60 |
| 48 x 96 | 700 | 1110 | 3.60 |
| 48 x 108 | 800 | 1230 | 3.68 |

Hot water storage tanks are ASME code constructed, galvanized carbon steel at 125 PSI working pressure. Inlet and outlet connections up to 6 inches are included. Manholes are included on 42-inch and larger diameter tanks.

The weight of factory installed tank heating coils is not included and therefore must be added to the above weight, dependent on size, for total lifting weight.

Manhours include receiving at job site, off-loading from carrier, moving within 50 feet of final location, rigging, picking, setting, and aligning.

Manhours do not include steam, condensate, or water piping, insulation, or instrumentation. See respective tables for these time frames.

# COMPRESSION TANKS

MANHOURS REQUIRED EACH

| Tank Size Inches | Capacity Gallons | Weight Pounds | Installation Manhours |
|---|---|---|---|
| 13 x 34-1/2 | 15 | 50 | 1.00 |
| 13 x 51 | 24 | 73 | 1.00 |
| 13 x 61 | 30 | 79 | 1.00 |
| 16-1/4 x 53 | 40 | 90 | 1.10 |
| 16-1/4 x 76-1/2 | 60 | 145 | 1.40 |
| 20-1/4 x 68 | 80 | 170 | 1.58 |
| 20-1/4 x 82 | 100 | 200 | 1.73 |
| 24-1/4 x 71-1/2 | 120 | 334 | 2.16 |
| 24-1/4 x 83-1/2 | 144 | 383 | 2.20 |
| 30 x 60 | 163 | 497 | 2.40 |

Compression tanks are ASME code construction, black carbon steel at 125 PSI working pressure and are prefabricated with gauge glass connections.

Manhours include receiving at job site, off-loading from carrier, moving within 50 feet of final location, rigging, picking, setting, and aligning.

Manhours do not include steam, condensate or water piping, insulation or instrumentation. See respective tables for these time requirements.

# COMBINATION HEATING & COOLING UNITS

MANHOURS REQUIRED EACH

| Furnace Capacity BTU/Hour | Gas BTU/Hour | Forced Air Fan Motor Horsepower | Cooling Capacity Tons | Nominal CFM | Unit Weight Pounds | Installation Manhours |
|---|---|---|---|---|---|---|
| 80,000 | 100,000 | 1/2 | 3 | 1,550 | 400 | 10.0 |
| 100,000 | 125,000 | 1/2 | 4 | 1,800 | 460 | 12.0 |
| 120,000 | 150,000 | 3/4 | 5 | 2,400 | 500 | 16.0 |

The above units include an upflow type gas-fired furnace and an electrically operated cooling system consisting of an evaporator coil factory mounted in bonnet of furnace, and a remote located, air-cooled condensing unit containing compressor, condenser coil, propeller fan and motor, and electrical controls. Cooling components are factory precharged with refrigerant and are equipped to receive furnished precharged tubing. Furnaces are certified for installation on combustible flooring.

Manhours include receiving at job site, off-loading from carrier, moving within 100 feet of final location, setting and aligning furnace, mounting air-cooled condenser a maximum of 100 feet from furnace, and installing a wall-mounted line voltage thermostat.

Manhours do not include installation of gas piping, power wiring or interconnecting control wiring. See respective tables for these time requirements.

# ROOF VENTILATORS— GRAVITY RELIEF TYPE

MANHOURS REQUIRED EACH

| CFM at 1/8 Inch Static | Roof Opening Inches | Weight Pounds | Installation Manhours |
|---|---|---|---|
| 715 | 10 x 10 | 30 | 1.50 |
| 1,130 | 14 x 14 | 40 | 1.50 |
| 2,020 | 18 x 18 | 95 | 2.00 |
| 3,130 | 24 x 24 | 115 | 2.00 |
| 5,030 | 30 x 30 | 185 | 2.25 |
| 6,930 | 36 x 36 | 250 | 3.00 |
| 11,600 | 48 x 48 | 275 | 3.00 |

# POWER EXHAUST TYPE

MANHOURS REQUIRED EACH

| CFM at 1/4 Inch Static | Horsepower | RPM | Weight Pounds | Installation Manhours |
|---|---|---|---|---|
| 225 | 1/25 | 1,550 | 25 | 1.50 |
| 465 | 1/15 | 1,550 | 30 | 1.50 |
| 620 | 1/8 | 1,140 | 50 | 1.50 |
| 1,250 | 1/4 | 1,725 | 50 | 1.50 |
| 1,500 | 1/3 | 1,725 | 65 | 1.75 |
| 1,890 | 1/2 | 1,725 | 100 | 2.00 |
| 2,650 | 1/2 | 1,140 | 150 | 2.00 |
| 3,600 | 3/4 | 1,140 | 175 | 2.25 |
| 5,950 | 1 | 680 | 300 | 3.00 |
| 6,900 | 1-1/2 | 775 | 305 | 3.00 |
| 8,570 | 2 | 725 | 325 | 3.00 |
| 11,300 | 3 | 680 | 400 | 4.00 |
| 12,300 | 3 | 620 | 410 | 4.00 |
| 16,900 | 3 | 390 | 925 | 5.00 |
| 20,600 | 5 | 460 | 955 | 5.00 |

Gravity relief vents are of the low contour type and include back-draft damper and bird screen.

Power exhausters are aluminum housing axial blade and includes automatic back-draft damper and motor.

Manhours include receiving and off-loading at job site, rigging, picking, setting, and aligning.

Manhours do not include cutting openings in roof, placement of curb caps and flashing, or electrical power hook-up. See respective tables for these time requirements.

# WALL EXHAUST FANS

MANHOURS REQUIRED EACH

| CFM at 0 Inch Static | Drive | Horsepower | Blade Size Inches | Installation Manhours |
|---|---|---|---|---|
| 200 | Direct | 1/100 | 8 | 2.50 |
| 325 | Direct | 1/70 | 8 | 2.50 |
| 360 | Direct | 1/70 | 10 | 2.60 |
| 530 | Direct | 1/40 | 10 | 2.60 |
| 875 | Direct | 1/20 | 12 | 2.75 |
| 1,480 | Direct | 1/20 | 14 | 2.85 |
| 1,950 | Direct | 1/20 | 16 | 2.90 |
| 3,080 | Direct | 1/4 | 18 | 2.95 |
| 3,500 | Direct | 1/4 | 20 | 3.00 |
| 4,000 | Direct | 1/4 | 24 | 3.50 |
| 6,800 | Belt | 3/4 | 24 | 4.00 |
| 20,000 | Belt | 3 | 36 | 5.50 |
| 31,000 | Belt | 5 | 42 | 7.00 |
| 36,000 | Belt | 7-1/2 | 48 | 7.50 |

The above are propeller, motor driven package units and include automatic back-draft shutter with RPM range of 1,500 to 1,750.

Manhours include receiving and off-loading at job site, moving within 50 feet of final location, and setting and aligning in an existing wall opening to 12 feet above floor.

Manhours do not include preparation of opening, or electrical hook-up. See respective tables for these time frames.

# HEAVY-GAUGE CENTRIFUGAL FANS

MANHOURS REQUIRED EACH

| Diameter of Wheel Inches | Maximum CFM Range | Approximate Weight Pounds | Installation Manhours |
|---|---|---|---|
| 12-1/4 | 2,100 | 140 | 4.40 |
| 13-1/2 | 2,700 | 165 | 4.50 |
| 15 | 2,900 | 185 | 5.70 |
| 16-1/2 | 3,500 | 195 | 6.00 |
| 18-1/4 | 5,000 | 235 | 6.40 |
| 20 | 6,500 | 300 | 6.50 |
| 22-1/4 | 8,000 | 330 | 7.30 |
| 24-1/2 | 9,100 | 425 | 7.40 |
| 27-1/2 | 12,000 | 525 | 7.55 |
| 30 | 14,750 | 610 | 8.00 |
| 33 | 18,000 | 775 | 8.40 |
| 36-1/2 | 25,000 | 940 | 9.60 |
| 40-1/4 | 30,000 | 1,575 | 12.00 |
| 44-1/2 | 37,000 | 1,870 | 13.80 |
| 49 | 45,000 | 2,225 | 16.80 |
| 54-1/4 | 55,000 | 2,750 | 22.60 |
| 60 | 67,000 | 3,050 | 30.00 |
| 66 | 81,000 | 3,900 | 43.40 |
| 73 | 99,000 | 5,625 | 56.50 |
| 80-3/4 | 122,000 | 6,850 | 70.70 |
| 89 | 148,000 | 8,800 | 82.80 |

Manhours include receiving and unloading at job site, moving within 50 feet of erection site, setting and aligning at floor level, and adjusting bearings.

Manhours do not include installation of inlet vane control, out dampers, motor and drives. See respective tables for these time frames.

# FAN MOTORS & V-BELT DRIVES

MANHOURS REQUIRED FOR ITEMS LISTED

| Motor Horsepower | Motor Weight Pounds | INSTALLATION MANHOURS | |
|---|---|---|---|
| | | Fan Motors | V-Belt Drives |
| 3 | 85 | 3.56 | 1.00 |
| 5 | 95 | 4.28 | 1.50 |
| 7-1/2 | 125 | 4.99 | 2.00 |
| 10 | 175 | 6.41 | 2.20 |
| 15 | 230 | 7.84 | 2.40 |
| 20 | 350 | 9.98 | 2.60 |
| 25 | 465 | 11.40 | 3.00 |
| 30 | 510 | 14.25 | 4.50 |
| 40 | 600 | 18.53 | 4.80 |
| 50 | 670 | 21.38 | 6.00 |
| 60 | 745 | 24.23 | 7.40 |
| 75 | 820 | 28.50 | 9.00 |
| 100 | 875 | 34.20 | – |

Fan motor manhours include mounting of motor on base and adjusting drive alignment.

V-belt manhours include installation of belt and alignment of wheels.

All manhours include receiving and off-loading at job site, uncrating, and moving within 50 feet of erection site.

Manhours do not include electrical hook-up. See respective table for this time requirement.

# CHILLED WATER PUMPS

MANHOURS REQUIRED EACH

| Pump GPM | Head Feet | Discharge Pipe Size | Motor Horsepower | Weight Pounds | Installation Manhours |
|---|---|---|---|---|---|
| 300 | 46 | 2-1/2″ | 5 | 340 | 12.0 |
| 480 | 87 | 3″ | 15 | 545 | 16.9 |
| 1,200 | 58 | 6″ | 25 | 1,450 | 20.2 |
| 1,800 | 46 | 6″ | 30 | 1,500 | 21.1 |
| 2,400 | 46 | 8″ | 40 | 2,000 | 23.9 |
| 3,000 | 46 | 8″ | 50 | 2,025 | 24.6 |
| 3,750 | 55 | 8″ | 75 | 2,200 | 27.6 |
| 4,200 | 66 | 8″ | 100 | 2,300 | 30.0 |
| 4,800 | 58 | 10″ | 100 | 2,450 | 30.6 |
| 5,500 | 50 | 12″ | 100 | 2,800 | 30.6 |

The pumps are cast iron, single-stage, horizontal cradle-mounted, vertical split case, end suction, top discharge, 125-pound flat face flanged with dripproof type motors, 1,700 RPM, direct connected by use of a flexible coupling. Pump and motor are both mounted on a common base plate.

Manhours include job handling, hauling, rigging, and setting and aligning of pump and motor.

Manhours do not include installation of piping, instrumentation and electrical motor starter and hook-up. See respective tables for these time frames.

# CONDENSER WATER PUMPS

MANHOURS REQUIRED EACH

| Pump GPM | Head Feet | Discharge Pipe Size | Motor Horsepower | Weight Pounds | Installation Manhours |
|---|---|---|---|---|---|
| 350 | 59 | 3″ | 7.5 | 480 | 13.8 |
| 600 | 69 | 4″ | 15 | 680 | 16.9 |
| 1,500 | 46 | 6″ | 25 | 1,450 | 20.2 |
| 2,200 | 50 | 8″ | 40 | 2,000 | 23.9 |
| 3,000 | 46 | 8″ | 50 | 2,025 | 24.6 |
| 3,700 | 56 | 8″ | 75 | 2,200 | 27.6 |
| 4,500 | 62 | 10″ | 100 | 2,450 | 30.6 |
| 5,250 | 53 | 12″ | 100 | 2,800 | 30.6 |
| 5,750 | 48 | 14″ | 100 | 4,000 | 35.0 |
| 6,000 | 46 | 14″ | 100 | 4,000 | 35.0 |

Pumps are cast iron, single-stage, horizontal cradle-mounted, vertical split case, end suction, top discharge, 125-pound flat face flanged with dripproof type motors, 1,750 RPM through 12-inch and 1,160 RPM for 14-inch. Pump and motor and both mounted on a common base plate.

Manhours include job handling, hauling, rigging, and setting and aligning of pump and motor.

Manhours do not include installation of piping, instruments and electrical motor starter, and hook-up. See respective tables for these time frames.

# CIRCULATING-BOOSTER PUMPS

MANHOURS REQUIRED EACH

| Pump GPM | Head Feet | Pipe Size Connections | Motor Horsepower | Weight Pounds | Installation Manhours |
|---|---|---|---|---|---|
| Standard | | | | | |
| 15 | 6 | 3/4″ | 1/8 | 28 | 10.0 |
| 20 | 5.5 | 1″ | 1/8 | 28 | 10.0 |
| 25 | 4.5 | 1-1/4″ | 1/8 | 28 | 10.0 |
| 30 | 2 | 1-1/2″ | 1/8 | 28 | 10.0 |
| High Velocity | | | | | |
| 30 | 7 | 1″ | 1/8 | 43 | 10.0 |
| 35 | 4 | 1-1/4″ | 1/8 | 43 | 10.0 |
| 35 | 5 | 1-1/2″ | 1/8 | 43 | 10.0 |
| High Head | | | | | |
| 10 | 9 | 1-1/2″ | 1/8 | 51 | 10.0 |
| 10 | 10 | 2″ | 1/6 | 42 | 10.0 |
| 10 | 16 | 1″ | 1/3 | 43 | 10.0 |

The pumps are of all bronze construction and are in-the-line, centrifugal type. Dripproof type motors are integral direct connected to pumps, 1,750 RPM, 115-volt, 60 Hz., single-phase. Pipe connections have threaded companion flanges bolted on.

Manhours include job handling, hauling, rigging, setting and aligning of pump and motor.

Manhours do not include installation of piping, instrumentation, and electrical motor starter, and hook-up. See respective tables for these time frames.

# AIR BALANCING & SYSTEM TESTING

MANHOURS REQUIRED EACH

| System | Balance & Test Manhours |
|---|---|
| Air Handlers–Water Side | 8.0 |
| Air Handlers–Air Side | 3.0 |
| Chiller Equipment | 5.0 |
| Wall Grills and Registers | 0.5 |
| Ceiling Air Diffusers | 0.8 |
| Roof Exhausters | 1.0 |
| Blowers to 5000 CFM | 2.0 |
| Blowers 5000 to 30,000 CFM | 3.5 |
| Induction and Fan Coil Units–Floor Mounted | 2.0 |
| Induction and Fan Coil Units–Ceiling Mounted | 3.0 |
| Smoke Stack Combustion Test | 3.0 |

Manhours include all operations to balance and test the systems in accordance with procedures which, in most cases, have been adapted from the Associated Air Balance Council.

# TEST & ADJUST CONTROLS

MANHOURS REQUIRED EACH

| Item | Manhours Each |
|---|---|
| Thermostat–Two Position | 0.5 |
| Thermostat–Modulating | 0.9 |
| Humidistat–Two Position | 0.9 |
| Humidistat–Modulating | 1.4 |
| Automatic Pump Down | 1.5 |
| Motor Control | 1.0 |
| Back-Pressure Control | 1.3 |
| Water Solenoid Valve | 0.4 |
| Water Regulating Valve | 0.8 |
| Expansion Valve | 0.9 |

Manhours include checking all terminals for tightness, testing for cycle in all positions, and making necessary adjustments.

Manhours for system and controls do not include installation of these items. See respective tables for these time frames.

# UNLOADING EQUIPMENT & TANKS FROM OPEN CARRIER

NET MANHOURS PER TON

| Machinery Classification | MANHOURS | | |
|---|---|---|---|
| | Group One | Group Two | Group Three |
| Lightweight & Bulky | | | |
| Up to 1500 lbs. | 2.80 | 3.85 | 3.15 |
| Lightweight & Easily Handled | | | |
| Up to 1500 lbs. | 2.10 | 3.15 | 2.45 |
| Heavyweight & Easily Handled | | | |
| Up to 10 tons | 1.75 | 2.10 | 1.90 |
| Up to 50 tons | 1.05 | 1.60 | 1.25 |
| Heavyweight & Bulky | | | |
| Up to 10 tons | 1.75 | 2.45 | 1.95 |
| Up to 50 tons | 1.40 | 2.30 | 1.75 |

*All groups:* Using forklift truck, derrick, crane or gin pole.

*Group One:* Unload to temporary storage adjacent to the carrier.

*Group Two:* Direct to floor location—second or third floor.

*Group Three:* To existing foundation or structural frame work.

All Groups include allowance for equipment operating crews.

# UNLOADING EQUIPMENT & TANKS FROM ENCLOSED CARRIER WITH END OR SIDE OPENING

NET MANHOURS PER TON

| Machinery Classification | MANHOURS | | |
|---|---|---|---|
| | Group One | Group Two | Group Three |
| Lightweight & Bulky | | | |
| (Up to 1500 lbs.) | 3.50 | 6.50 | 4.90 |
| Lightweight & Easily Handled | | | |
| (Up to 1500 lbs.) | 2.65 | 3.50 | 3.35 |
| Heavyweight & Easily Handled | | | |
| (Up to 10 Tons) | 2.80 | – | 3.70 |
| (Up to 50 Tons) | 2.10 | – | 3.00 |
| Heavyweight & Bulky | | | |
| (Up to 10 Tons) | 3.15 | – | 4.20 |
| (Up to 50 Tons) | 2.80 | – | 3.75 |

*Group One:* Using fork truck or other power equipment, drag object from inside to opening and unload to ground or location if adjacent.

*Group Two and Three:* Jacking, bulling, skidding on small rollers, then drag to opening.

*Group Two:* Remove to ground by hand or slide.

*Group Three:* Remove to ground by power equipment.

All Groups include allowance for equipment operating crews.

# HANDLING AND HAULING EQUIPMENT AND TANKS

**Move Manually With Some or All of the Individual Specified**

NET MANHOURS PER TON

| Operation | MACHINERY CLASSIFICATION | | | | | |
|---|---|---|---|---|---|---|
| | Lt. Wt. & Bulky | Lt. Wt. & Easily Handled | Hvy. Wt. & Easily Handled | | Hvy. Wt. & Bulky | |
| | To 1500# | To 1500# | To 10 tons | To 50 tons | To 10 tons | To 50 tons |
| Jacking up & placing rollers | 0.70 | 0.56 | 0.46 | 0.35 | 0.53 | 0.46 |
| Moving on skids or small rollers for 100 feet | 1.05 | 0.98 | 0.42 | 0.35 | 0.60 | 0.53 |
| Jacking up or down, placing or removing cribbing per ft. of height | 0.70 | 0.56 | 0.46 | 0.35 | 0.53 | 0.42 |
| Bulling and moving or turning up to 10 feet. | 1.40 | 1.23 | 0.88 | 0.70 | 0.95 | 0.81 |
| Handling cribbing & timber per piece | 0.04 | 0.04 | 0.04 | 0.04 | 0.04 | 0.04 |

# HANDLING AND HAULING EQUIPMENT AND TANKS

**Move by Ford Truck, Crane, Hand Truck, Dolly Truck**

NET MANHOURS PER TON

| Operation | MACHINERY CLASSIFICATION | | | | | |
|---|---|---|---|---|---|---|
| | Lt. Wt.& Bulky To 1500 lbs. | Lt. Wt. & Easily Handled To 1500 lbs. | Hvy. Wt. & Easily Handled | | Hvy. Wt. & Bulky | |
| | | | To 10 Tons | To 50 Tons | To 10 Tons | To 50 Tons |
| Transport for 100 ft. including one lifting operation | 0.40 | 0.35 | 0.33 | 0.29 | 0.38 | 0.31 |
| Place on base as part of transporting. Does not include line-up. | 0.35 | 0.32 | 0.22 | 0.18 | 0.23 | 0.20 |
| Build up cribbins-set object on top prior to lowering or horizontal positioning. | 0.77 | 0.63 | 0.53 | 0.42 | 0.60 | 0.49 |

All groups include allowance for equipment operation crews.

# ALIGNMENT OF EQUIPMENT

MANHOURS PER WEIGHT UNITS LISTED

| Machinery Classification In Pounds | MANHOURS | | | |
|---|---|---|---|---|
| | Group One | Group Two | Group Three | Group Four |
| 200 or less | 0.60 | 2.31 | 3.47 | 1.40 |
| 500 | 0.74 | 2.94 | 4.41 | 1.40 |
| 750 | 0.81 | 3.22 | 4.83 | 1.40 |
| 1000 | 0.91 | 3.57 | 5.36 | 1.40 |
| 1500 | 1.05 | 4.20 | 6.30 | 1.40 |
| 2000 | 1.23 | 4.97 | 6.92 | 1.40 |
| 2500 | 1.47 | 5.95 | 7.65 | 1.40 |
| 3000 | 1.86 | 6.30 | 9.45 | 1.40 |
| 4000 | 2.63 | 8.70 | 10.88 | 1.40 |
| Per ton above 2 | 1.40 | 5.20 | 7.20 | 1.40 |

*Group One:* Rough alignment if a separate operation. Setting, raising or lowering and removing temporary supporting timbers if used.

*Group Two:* Accurate alignment - pre-assembled at vendors shops, delivered as a single unit. Rough alignment included if combined operations.

*Group Three:* Accurate alignment - disassembled into major sections. Reassembled on frame at final location.

*Group Four:* Grouting per square foot.

Allowance for equipment operating crews included.

# FABRICATION AND INSTALLATION OF SAFETY V-BELT MOTOR GUARDS

NET MANHOURS PER UNIT LISTED

| Motor Horsepower | 18 Gauge Sheet Metal Pounds | Miscellaneous Steel Pounds | MANHOURS | | |
|---|---|---|---|---|---|
| | | | Shop Fabricated | Field Erection | Total |
| 2 or less | 28 | 22 | 2.0 | 2.1 | 4.1 |
| 5 | 69 | 33 | 4.0 | 4.2 | 8.2 |
| 7-1/2 | 93 | 40 | 5.0 | 5.3 | 10.3 |
| 10 | 105 | 45 | 5.0 | 5.3 | 10.3 |
| 15 | 145 | 47 | 7.5 | 7.9 | 15.4 |
| 20 | 170 | 51 | 8.0 | 8.4 | 16.4 |
| 25 | 215 | 58 | 9.6 | 10.1 | 19.7 |
| 30 | 245 | 58 | 10.4 | 10.9 | 21.3 |
| 40 | 290 | 66 | 12.8 | 13.4 | 26.2 |
| 50 | 325 | 69 | 14.4 | 15.1 | 29.5 |
| 75 | 430 | 77 | 17.6 | 18.5 | 36.1 |
| 100 | 505 | 83 | 20.3 | 21.3 | 41.6 |
| 150 | 760 | 115 | 29.3 | 30.8 | 60.1 |
| 200 | 990 | 165 | 40.0 | 42.0 | 82.0 |

Manhours include handling, hauling, fabrication and installing safety V-belt motor guards as outlined above.

Manhours do not include motor installations. See respective table for this charge.

# ARC PLATE BUTT WELDING

NET MANHOURS PER LINEAR FOOT

| | Plate Thickness Inches | | | | | | |
|---|---|---|---|---|---|---|---|
| | 1/8 | 3/16 | 1/4 | 5/16 | 3/8 | 7/16 | 1/2 |
| Butt Weld | | | | | | | |
| Flat | – | – | .24 | .30 | .38 | .38 | .44 |
| Vertical | – | – | .29 | .37 | .51 | .51 | .58 |
| Horizontal | – | – | .35 | .42 | .64 | .64 | .71 |
| Overhead | – | – | .37 | .45 | .77 | .77 | .84 |
| Flame Cutting | – | .09 | .09 | .09 | .10 | .10 | .11 |
| | Plate Thickness Inches | | | | | | |
| | 9/16 | 5/8 | 3/4 | 7/8 | 1 | 1-1/8 | 1-1/4 |
| Butt Weld | | | | | | | |
| Flat | – | .58 | .70 | .77 | .83 | 1.01 | 1.12 |
| Vertical | – | .84 | .88 | 1.00 | 1.07 | 1.31 | 1.41 |
| Horizontal | – | 1.07 | 1.07 | 1.12 | 1.17 | 1.40 | 1.49 |
| Overhead | – | 1.17 | 1.17 | 1.31 | 1.35 | 1.76 | 1.82 |
| Flame Cutting | .11 | .12 | .13 | .16 | .17 | .18 | .20 |

Above manhours include welder and helper time necessary for set-up of machine, procuring welding materials, tackwelding when necessary and welding.

Manhours are based on 100 linear feet or more of welding of the type and size listed. If less than 100 linear feet of welding is required, manhours should be increased by at least 25 per cent.

Manhours do not include setting, aligning or positioning of plate or scaffolding. See respective tables for these charges.

# ARC PLATE FILLET WELDING

NET MANHOURS PER LINEAR FOOT

| | Plate Thickness Inches | | | | | | |
|---|---|---|---|---|---|---|---|
| | 1/8 | 3/16 | 1/4 | 5/16 | 3/8 | 7/16 | 1/2 |
| **Fillet Weld** | | | | | | | |
| Flat | .08 | .10 | .17 | .20 | .26 | .29 | .34 |
| Vertical | .10 | .14 | .20 | .27 | .34 | .36 | .42 |
| Horizontal | .11 | .15 | .18 | .32 | .39 | .40 | .46 |
| Overhead | .14 | .17 | .21 | .37 | .42 | .43 | .50 |
| Flame Cutting | – | .09 | .09 | .09 | .10 | .10 | .11 |
| | **Plate Thickness Inches** | | | | | | |
| | 9/16 | 5/8 | 3/4 | 7/8 | 1 | 1-1/8 | 1-1/4 |
| **Fillet Weld** | | | | | | | |
| Flat | .38 | .45 | .50 | .58 | .64 | – | – |
| Vertical | .45 | .53 | .64 | .71 | .77 | – | – |
| Horizontal | .50 | .58 | .71 | .77 | .83 | – | – |
| Overhead | .54 | .64 | .77 | .83 | .89 | – | – |
| Flame Cutting | .11 | .12 | .13 | .16 | .17 | .18 | .20 |

Above manhours include welder and helper time necessary for set-up of machine, procuring welding materials, tackwelding when necessary and welding.

Manhours are based on 100 linear feet or more of welding of the type and size listed. If less than 100 linear feet of welding is required, manhours should be increased by at least 25 per cent.

Manhours do not include setting, aligning or positioning plate or scaffolding. See respective tables for these charges.

# Section 4

# INSULATION AND WATERPROOFING

The following tables cover the installation of waterproofing and insulation for heating, ventilating, and air conditioning systems within a process or industrial plant. Because there are many types and applications of insulation, the manhour tables in this section are those that are more commonly used for this type work.

Membrane waterproofing, dampproofing, and insulmastic applications are usually subcontracted to a contractor who specializes in this type of work. However, manhour tables have been included for these applications in order to assist the estimator should a company desire to perform these operations with their own forces.

# DUCT FINISHES

NET MANHOURS PER SQUARE FOOT

| Type | Manhours |
|---|---|
| Two coats cement plaster with wire mesh on light pasted canvas | .075 |
| Asphalt-emulsion finish | .068 |
| Sewed on 8 oz. canvas | .023 |
| Outdoor weatherproofing | .027 |
| Mastic over cement waterproofing | .014 |
| One coat 1/2-inch cement | .038 |
| Two coats paint on muslin or canvas | .020 |

Manhours include all necessary labor to complete above operations.

Manhours do not include scaffolding. See respective tables for this charge.

# DUCT INSULATION

NET MANHOURS PER SQUARE FOOT

| Type | Thickness Inches | Manhours |
|---|---|---|
| Wired on cork or rock cork | 1 | .113 |
| Wired on cork or rock cork | 2 | .120 |
| Fiberglas – standard vapor seal | | |
| 3 pounds per cubic foot | 1 | .120 |
| 6 pounds per cubic foot | 1 | .120 |
| Air-cell board or flexible fiberglas (concealed) | 1 | .097 |
| Celotex | 1 | .113 |

Manhours include handling, hauling and placing of insulating materials as outlined above.

Manhours do not include scaffolding. See respective table for this charge.

# HOT PIPING INSULATION

NET MANHOURS

| Pipe Size | Thickness & Type | Straight Pipe Per Lin. Ft. | Screwed & Weld Ftgs. Per Ea. | Flanges Per Pr. | Flanged Valves & Ftgs. Ea. |
|---|---|---|---|---|---|
| 1/2 | 1" thick Calsilite | .11 | .14 | .36 | .74 |
| 3/4 | 1" thick Calsilite | .11 | .15 | .36 | .74 |
| 1 | 1" thick Calsilite | .12 | .18 | .36 | .74 |
| 1-1/2 | 1" thick Calsilite | .13 | .21 | .41 | .83 |
| 2 | 1" thick Calsilite | .14 | .22 | .44 | .88 |
| 3 | 1" thick Calsilite | .18 | .27 | .54 | 1.37 |
| 4 | 1" thick Calsilite | .21 | .34 | .65 | 1.65 |
| 5 | 1" thick Calsilite | .25 | .52 | .72 | 2.05 |
| 6 | 1" thick Calsilite | .25 | .61 | .77 | 2.15 |
| 7 | 1½" thick Calsilite | .33 | .93 | .96 | 2.96 |
| 8 | 1½" thick Calsilite | .36 | 1.18 | 1.10 | 3.39 |

Thicknesses and manhours are for all hot services, if calcium silicate is used.

Manhours are for either indoor or outdoor service.

Bent Pipe – 1.5 x straight pipe of like size and thickness measured along outside radius.

Steam Traced Piping – to be manhoured at size of pipe covering required to fit over pipe and tracer line.

Method of Measurement – Straight pipe to be determined by measuring along approximate center line over the exterior of the insulation from center line to center line of change of direction. Measurement shall be made through all valves and fittings, except bent pipe.

Specifications:

a. Pipe covering – molded sections secured with 16 gauge galvanized tie wire. Finish – indoors with 6-ounce canvas with laps sealed with Arabol lagging adhesive. Finish – outdoor with 55# Fiberrock Asbestos Roofing felt secured with 16 gauge galvanized tie wire 6″ O.C.

b. Fittings – to be built-up with insulation cement or sectional pipe covering pointed up with asbestos cement, finished with 6-ounce canvas and Arabel for indoor service and "Seal Perm" for outdoor service.

# PIPE INSULATION—INDOOR THERMAL TYPE

NET MANHOURS

| Thickness Inches | Pipe Size | Straight Pipe per LF | Bent Pipe per LF | Flanges Line per Ea. | Valves Flgd. per Ea. | Valves S & W per Ea. | Fittings Flanged per Ea. | Fittings S & W per Ea. | Hangers Pipe per Ea. | Nozzles per Each |
|---|---|---|---|---|---|---|---|---|---|---|
| 1.0 | 1/2 | .18 | .28 | .56 | 1.50 | .75 | 1.50 | .28 | .18 | .18 |
| | 3/4 | .19 | .29 | .59 | 1.58 | .79 | 1.58 | .29 | .19 | .19 |
| | 1 | .21 | .31 | .63 | 1.69 | .84 | 1.69 | .31 | .21 | .21 |
| | 1-1/2 | .24 | .36 | .72 | 1.92 | .96 | 1.92 | .36 | .24 | .24 |
| | 2 | .25 | .38 | .76 | 2.04 | 1.02 | 2.04 | .38 | .25 | .25 |
| | 3 | .31 | .47 | .94 | 2.52 | 1.26 | 2.52 | .47 | .31 | .31 |
| | 4 | .37 | .56 | 1.12 | 2.99 | 1.49 | 2.99 | .74 | .37 | .37 |
| | 6 | .43 | .64 | 1.29 | 3.45 | 1.72 | 3.45 | .86 | .43 | .43 |
| 1.5 | 1/2 | .28 | .43 | .86 | 2.30 | 1.15 | 2.30 | .43 | .28 | .28 |
| | 3/4 | .30 | .45 | .90 | 2.42 | 1.21 | 2.42 | .45 | .30 | .30 |
| | 1 | .31 | .47 | .95 | 2.54 | 1.27 | 2.54 | .47 | .31 | .31 |
| | 1-1/2 | .35 | .53 | 1.06 | 2.84 | 1.42 | 2.84 | .53 | .35 | .35 |
| | 2 | .37 | .56 | 1.13 | 3.01 | 1.50 | 3.01 | .56 | .37 | .37 |
| | 3 | .44 | .66 | 1.34 | 3.57 | 1.78 | 3.57 | .66 | .44 | .44 |
| | 4 | .50 | .76 | 1.52 | 4.06 | 2.03 | 4.06 | 1.01 | .50 | .50 |
| | 6 | .57 | .86 | 1.73 | 4.63 | 2.31 | 4.63 | 1.15 | .57 | .57 |
| | 8 | .67 | 1.01 | 2.03 | 5.43 | 2.71 | 5.43 | 1.69 | .67 | .67 |
| | 10 | .80 | 1.21 | 2.43 | 6.48 | 3.24 | 6.48 | 2.02 | .80 | .80 |
| | 12 | .91 | 1.36 | 2.73 | 7.30 | 3.65 | 7.30 | 2.73 | .91 | .91 |
| | 14 | 1.01 | 1.52 | 3.05 | 8.14 | 4.07 | 8.14 | 3.05 | 1.01 | 1.01 |
| | 16 | 1.14 | 1.71 | 3.43 | 9.15 | 4.57 | 9.15 | 4.56 | 1.14 | 1.14 |
| | 18 | 1.27 | 1.90 | 3.80 | 10.17 | 5.08 | 10.17 | 6.35 | 1.27 | 1.27 |
| | 20 | 1.39 | 2.08 | 4.17 | 11.13 | 5.56 | 11.13 | 6.94 | 1.39 | 1.39 |
| | 24 | 1.62 | 2.43 | 4.87 | 12.99 | 6.49 | 12.99 | 9.74 | 1.62 | 1.62 |
| 2.5 | 1/2 | .47 | .71 | 1.42 | 3.79 | 1.89 | 3.79 | .71 | .47 | .47 |
| | 3/4 | .48 | .72 | 1.45 | 3.88 | 1.94 | 3.88 | .72 | .48 | .48 |
| | 1 | .50 | .76 | 1.52 | 4.06 | 2.03 | 4.06 | .76 | .50 | .50 |
| | 1-1/2 | .55 | .82 | 1.65 | 4.41 | 2.20 | 4.41 | .82 | .55 | .55 |
| | 2 | .58 | .87 | 1.74 | 4.65 | 2.32 | 4.65 | .87 | .58 | .58 |
| | 3 | .68 | 1.02 | 2.04 | 5.30 | 2.72 | 5.30 | 1.02 | .68 | .68 |
| | 4 | .78 | 1.16 | 2.33 | 6.21 | 3.10 | 6.21 | 1.55 | .78 | .78 |
| | 6 | .86 | 1.28 | 2.58 | 6.88 | 3.44 | 6.88 | 1.72 | .86 | .86 |
| | 8 | .97 | 1.46 | 2.93 | 7.81 | 3.90 | 7.81 | 2.43 | .97 | .97 |
| 3.5 | 1/2 | .74 | 1.12 | 2.24 | 6.00 | 3.00 | 6.00 | 1.12 | .74 | .74 |
| | 3/4 | .78 | 1.18 | 2.36 | 6.31 | 3.15 | 6.31 | 1.18 | .78 | .78 |
| | 1 | .80 | 1.20 | 2.40 | 6.42 | 3.21 | 6.42 | 1.20 | .80 | .80 |
| | 1-1/2 | .86 | 1.29 | 2.59 | 6.91 | 3.45 | 6.91 | 1.29 | .86 | .86 |
| | 2 | .91 | 1.37 | 2.74 | 7.32 | 3.66 | 7.32 | 1.37 | .91 | .91 |
| | 3 | 1.02 | 1.54 | 3.08 | 8.22 | 4.11 | 8.22 | 1.54 | 1.02 | 1.02 |
| | 4 | 1.11 | 1.67 | 3.34 | 8.93 | 4.46 | 8.93 | 2.23 | 1.11 | 1.11 |
| | 6 | 1.21 | 1.81 | 3.63 | 9.69 | 4.84 | 9.69 | 2.42 | 1.21 | 1.21 |
| | 8 | 1.35 | 2.03 | 4.06 | 10.84 | 5.42 | 10.84 | 3.38 | 1.35 | 1.35 |

*Thermal Insulation:* Consists of applying hydraulic setting, insulating cement by spraying, brushing, troweling or palming, coating with vinyl emulsion, double wrapping with glass fiber cloth, and coating with vinyl emulsion seal coat.

*Outside Use:* Add 10% to above manhours.

*Foamglass:* Use same manhours as appear above for this type insulation. This will include labor for butter joints with "Seal Koat" and secure with 16 and 14 gauge galvanized wire on 9" centers. Finish with one coat "Seal Koat" for indoor piping and 55# asbestos roofing felt secured with 16 gauge wire 6" on center over the layer of "Seal Koat" on outside piping.

*Note: S & W denotes screwed and welded.*

# INSULATION OF VESSELS, TANKS AND HEAT EXCHANGERS

NET MANHOURS PER SQUARE FOOT

| Type | Manhours |
|---|---|
| One layer blocks (wired on) | .048 |
| Additional layer of blocks (wired on) | .032 |
| Sponge Felt | .040 |
| Wire Mesh | .008 |
| Cement 1/4-Inch Thick | .024 |
| Cement 1/2-Inch Thick | .032 |
| Sewed on 8-oz. Canvas | .032 |
| Pasted on 8-oz. Canvas | .024 |
| Metal Lathe | .036 |
| Rosin Paper | .006 |
| 1-inch Thick Hair Felt | .024 |
| One Layer Asbestos Paper (1/32-inch thick) | .006 |
| Hot Pitch or Asphalt - One Mopping | .007 |
| Two Coats of Lead and Oil Paints | .018 |
| Concrete Primer and Enamel | .034 |
| Cold Water Paint | .008 |
| Add for Weatherproofing with Standard Covers | .024 |

Manhours include all operations necessary for the complete installation of the type insulation as outlined above.

Manhours do not include equipment installation or scaffolding, See respective tables for these charges.

# MEMBRANE WATERPROOFING VESSELS, TANKS AND HEAT EXCHANGERS

NET MANHOURS PER SQUARE FOOT

| Item | Manhours |
|---|---|
| One-ply Fabric and Two Moppings | .03 |
| Two-ply Fabric and Three Moppings | .05 |
| Three-ply Fabric and Four Moppings | .06 |
| Four-ply Fabric and Five Moppings | .07 |
| Each Additional Layer of Felt and Mopping | .02 |

Manhours are for the waterproofing and dampproofing as itemized and outlined above and include all labor operations necessary for this type of work.

Manhours do not include installation of equipment or scaffolding. See respective tables for these charges.

# WATERPROOFING, DAMPPROOFING & INSULMASTIC

MANHOURS PER SQUARE FOOT

| Item | Manhours | | | |
|---|---|---|---|---|
| | Cement Finisher | Roofer | Insulator | Total |
| Foundation & Basement Walls | | | | |
| Mastic type coating - 1/8" | .050 | – | – | .050 |
| Silicone or metallic coating | .020 | – | – | .020 |
| Damp resisting paint | .010 | – | – | .010 |
| Membrane & pitch (1 ply - 2 moppings) | – | .030 | – | .030 |
| Floors - Membrane & Pitch | | | | |
| Dry floor (2 ply - 3 moppings) | – | .042 | – | .042 |
| Damp floor (2 ply - 3 moppings) | – | .067 | – | .067 |
| Application of Insulmastic | | | | |
| Brushed on coat 1/16" | – | – | .020 | .020 |
| Brushed on coat 1/8" | – | – | .050 | .050 |
| Brushed on coat 1/4" | – | – | .068 | .068 |

Manhours are for the waterproofing and dampproofing as itemized above and include all labor operations as may be necessary for this type of work.

Manhours do not include the placement of concrete or concrete items. See respective tables for these charges.

# Section 5

# INSTRUMENTS AND CONTROLS

The manhour tables in this section are for the installation of instruments and controls commonly used in the heating, ventilating, air-conditioning, or mechanical systems of a process or industrial plant. The great variety of control combinations for any of these systems prevents complete coverage of all possibilities.

Packaged units ordered from a manufacturer should include factory-installed instruments and controls that may be required. This eliminates the excessive cost of field installation.

The manhours include time allowance to complete all operations as may be required for the installation of the described items in accordance with the notes that accompany each table.

# CONTROL BOARD & PANEL INSTRUMENTS

MANHOURS PER ITEMS LISTED

| Item | Installation Manhours |
|---|---|
| Control Board to 20 Feet Long | 26.0 |
| Differential Pressure Cell | 7.0 |
| Flow Instruments | 10.0 |
| Pressure Instruments | 8.0 |
| Liquid Level Instruments | 8.0 |
| Temperature Instruments | 10.0 |
| Air Filter | 1.0 |
| Air Regulator | 0.5 |
| Multipoint Recorder Receiver | 12.0 |

Instrument manhours include handling, hauling, unpacking, laying our on existing board, aligning, erecting, and adjusting of instruments to 2000 PSI.

Control board manhours include job receiving, handling, hauling, and installing control board.

Manhours do not include installation of pipe, valves, fitting, or electrical hook-up. See respective tables for these time requirements.

# THERMOSTATS

MANHOURS REQUIRED EACH

| Item and Description or Type | Installation Manhours |
|---|---|
| Line Voltage—Heavy Duty | |
| Heating—Bellows-Operated Microswitch | 1.5 |
| Cooling—Bellows-Operated Microswitch | 1.5 |
| Two-Stage—Bellows-Operated Mercury Bulb | 1.8 |
| Three Stage—Bellows-Operated Mercury Bulb | 2.0 |
| Motor Controller—Proportional (60 to 87°F) | 2.5 |
| Motor Controller—High or Low Limit | 2.5 |
| Low Temperature Remote Bulb (−20 to +50°F) | 2.2 |
| For Explosion Proof Add | 2.0 |
| Low Voltage—Light Duty | |
| Heating—Coiled Bimetal | 1.5 |
| Cooling—Coiled Bimetal | 1.5 |
| Heating and Cooling—Coiled bimetal | 1.7 |
| Multi-Stage, 1 Cool +1 Heat—Coiled Bimetal | 1.7 |
| Multi-Stage, 2 Cool +1 Heat—Coiled Bimetal | 1.9 |
| Multi-Stage, 1 Cool +2 Heat—Coiled Bimetal | 2.1 |
| Multi-Stage, 2 Cool or 2 Heat—Coiled Bimetal | 2.1 |
| Humidstat—Relative Humidity Range 20 to 80% | 1.5 |

Manhours include job handling, hauling, unpacking, installing, and hook-up.

Manhours do not include installation of electrical source. See respective tables for these time frames.

# ORIFICE PLATES

MANHOURS REQUIRED EACH

| Orifice Flange Size Inches | Installation Manhours |
|---|---|
| 1-1/2 | 1.0 |
| 2 | 1.2 |
| 3 | 2.0 |
| 4 | 2.9 |
| 6 | 3.7 |
| 8 | 4.9 |
| 10 | 6.3 |
| 12 | 7.7 |

Manhours include job handling, hauling, unbolting of orifice flanges, positioning the orifice plate, rebolting the flanges, testing and final checkout of 300-pound or 600-pound, 316 stainless steel paddle type orifice plates, 1/8 inch or 1/4 inch thick.

Manhours do not include installation of flanges. See piping section for these time frames.

# LEVEL GAUGE GLASSES

MANHOURS REQUIRED EACH

| Nominal Visible Length Inches | Installation Manhours |
|---|---|
| 6-3/4 | 3.6 |
| 10-1/4 | 4.0 |
| 13 | 4.6 |
| 19-3/4 | 5.0 |
| 26-3/4 | 5.3 |
| 33-3/4 | 5.7 |
| 45-1/2 | 6.4 |
| 55 | 7.4 |
| 65-3/8 | 7.7 |
| 78-3/4 | 8.5 |

Gauge glasses are transparent type with a rating of 2000 PSI at 100°F, and 1100 PSI at 750°F with carbon steel body screwed assembly.

Manhours include job handling, hauling, unpacking, and installing of gauge glass and gauge valve.

Manhours do not include piping, conduit, or wire installation. See respective tables for these time frames.

# FLOW VENTURI BALANCING DEVICES

MANHOURS REQUIRED EACH

| Pipe Size Inches | GPM Range | Installation Manhours |
|---|---|---|
| 1/2 | 0.2- 4.0 | 1.4 |
| 3/4 | 0.5- 6.0 | 1.6 |
| 1 | 2.0- 15.0 | 1.8 |
| 1-1/4 | 4.0- 23.0 | 1.9 |
| 1-1/2 | 6.0- 30.0 | 2.1 |
| 2 | 8.0- 50.0 | 2.2 |
| 2-1/2 | 12.0- 70.0 | 2.4 |
| 3 | 20.0- 100.0 | 2.8 |
| 4 | 30.0- 180.0 | 4.1 |
| 6 | 70.0- 400.0 | 5.7 |
| 8 | 150.0- 700.0 | 6.7 |
| 10 | 200.0-1,000.0 | 8.1 |
| 12 | 300.0-1,750.0 | 8.8 |

Venturis come complete with valves and quick-disconnect connections and are for a working pressure to 275 PSI at 100°F. Sizes 1/2-inch through 2-inch are brass with screwed joints and sizes 2-1/2-inch through 12-inch are cast iron or carbon steel weldneck 150-pound flanged.

Manhours include job handling, hauling, and total installation of balancing device.

Manhours do not include piping installation. See respective tables for these time frames.

# INSTALLATION OF CABLE TRAY & FITTINGS

MANHOURS PER UNITS LISTED

| Tray Item Description | Unit | Width of Tray | | | | | | |
|---|---|---|---|---|---|---|---|---|
| | | 6″ | 9″ | 12″ | 18″ | 24″ | 30″ | 36″ |
| Ladder Type Cable Tray–Straight | LF. | 0.25 | 0.30 | 0.33 | 0.35 | 0.40 | 0.45 | 0.55 |
| 90° Horizontal Elbows–12″ Radius | Ea. | 1.25 | 1.25 | 1.50 | 1.90 | 2.50 | 3.00 | 3.50 |
| 90° Vertical Elbows–12″ Radius | Ea. | 2.19 | 2.19 | 2.63 | 3.33 | 4.38 | 4.98 | 5.80 |
| Horizontal Tees–12″ Radius | Ea. | 2.30 | 2.30 | 2.75 | 3.50 | 4.60 | 5.25 | 6.10 |
| Horizontal Crosses–12″ Radius | Ea. | 3.00 | 3.00 | 3.60 | 4.55 | 6.00 | 6.85 | 7.95 |
| Reducer | Ea. | – | – | 3.00 | 3.50 | 4.00 | 4.50 | 5.00 |
| Expansion Joint | Ea. | 2.50 | 3.00 | 4.00 | 4.75 | 5.50 | 6.25 | 7.00 |
| Connector Plates | Pr. | 1.00 | 1.00 | 1.00 | 1.00 | 1.00 | 1.00 | 1.00 |
| Dropouts | Ea. | 1.25 | 1.25 | 1.50 | 1.75 | 2.00 | 2.50 | 3.00 |
| Blind Ends | Ea. | 0.50 | 0.50 | 1.00 | 1.00 | 1.25 | 1.50 | 1.75 |
| Tray Cover Plate | LF. | 0.10 | 0.12 | 0.15 | 0.20 | 0.25 | 0.50 | 0.75 |
| Cable Separators | Ea. | 1.00 | 1.00 | 1.00 | 1.00 | 1.00 | 1.00 | 1.00 |

Manhours are for installation of ladder type cable tray and fittings with 3-3/8-inch siderails and rungs on 6-inch centers all of 16 gauge steel.

Manhours include job handling, hauling, cutting, assembling, and placing.

Manhours do not include structural supports on which cable tray is installed.

# MISCELLANEOUS IN-LINE INSTRUMENTS

MANHOURS REQUIRED EACH

| Line Size | PRESSURE RATING | | | | | |
|---|---|---|---|---|---|---|
| | 150-Lb. | 300-Lb. | 400-Lb. | 600-Lb. | 900-Lb. | 1500-Lb. |
| 1 | 1.7 | 1.9 | – | 2.4 | – | 3.0 |
| 1-1/2 | 1.8 | 2.0 | – | 2.8 | – | 3.6 |
| 2 | 1.9 | 2.4 | – | 3.1 | – | 3.9 |
| 3 | 2.8 | 3.3 | – | 4.0 | 4.0 | 5.4 |
| 4 | 4.1 | 4.8 | 4.8 | 5.6 | 5.6 | 7.3 |
| 6 | 5.2 | 6.1 | 6.1 | 7.8 | 7.8 | 9.3 |
| 8 | 7.0 | 8.2 | 8.2 | 9.4 | 9.4 | 12.7 |
| 10 | 9.0 | 10.2 | 10.2 | 11.5 | 11.5 | – |
| 12 | 11.1 | 12.7 | 12.7 | 14.5 | 14.5 | – |
| 14 | 12.7 | 14.6 | 14.6 | – | – | – |
| 16 | 14.7 | 16.9 | 16.9 | – | – | – |
| 18 | 16.3 | 18.9 | 18.9 | – | – | – |
| 20 | 18.7 | 21.6 | 21.6 | – | – | – |
| 24 | 21.7 | 25.1 | – | – | – | – |
| 30 | 27.2 | – | – | – | – | – |
| 36 | 32.6 | – | – | – | – | – |

There are a variety of instruments that when installed, actually become a part of the pipeline. The manhours take into consideration the installation of the instrument including two flange-ups, checking out of storage, handling, calibrating when necessary, and testing.

Manhours do not include the installation of pipe, valves, fittings, conduit, wire, cable tray, or supports. See other tables for these time frames.

# MISCELLANEOUS TEMPERATURE, PRESSURE, AND OTHER INSTRUMENTS

MANHOURS REQUIRED EACH

| Instrument Description | Manhours |
|---|---|
| Temperature | |
| Bi-Metal Thermometer | 2.0 |
| Thermowell | 2.0 |
| Thermowell with Chain and Cap | 2.0 |
| Thermocouple Assembly–Screwed | 3.0 |
| Thermocouple Assembly–Flanged | 4.0 |
| Thermocouple Assembly–Flanged R.J. | 4.5 |
| Temperature Switch with Capillary Tube–Explosion Proof Mercoid | 3.5 |
| Pressure | |
| Pressure Gauge–4-1/2" Dial | 3.0 |
| Draft Gage | 4.0 |
| Pressure Switch–Explosion Proof | 3.0 |
| Miscellaneous | |
| Siphon | 0.5 |
| Valve Positioners | 2.5 |
| Integrally-Mounted Valve Positioner | 4.0 |
| Valve Operator | 5.0 |
| Air Filters | 1.5 |
| Air Pressure Regulator | 1.0 |
| Air Pressure Regulator (Combination) | 1.5 |
| Adjustable Restrictor (Damper) | 2.0 |
| Alarms (Panel-Mounted Single) | 1.5 |
| Alarms (Panel-Mounted Dual) | 2.0 |
| Vibration Switches–Electric | 4.0 |

Manhours include checking out of storage, handling, installing, and calibrating and testing of instrument where necessary.

Manhours do not include installation of pipe, valves, fittings, conduit, wire, cable tray, are supports. See other tables for these time frames.

# INSTALLATION OF MULTI-TUBE BUNDLES

MANHOURS PER LINEAR FOOT

| Tube Type | Size inches | Number of Tubes in Bundle 2 | 3 | 4 | 5 | 7 | 8 | 10 | 12 | 14 | 19 | 37 |
|---|---|---|---|---|---|---|---|---|---|---|---|---|
| Copper with | 1/4 | .04 | .05 | .06 | .07 | .09 | .10 | .12 | .14 | .16 | .22 | .32 |
| Plastic Sheath | 3/8 | .05 | .06 | .07 | .08 | .11 | .14 | .18 | .22 | – | – | – |
| Only | 1/2 | .07 | .09 | .13 | – | – | – | – | – | – | – | – |
| Copper with | 1/4 | .05 | .06 | .07 | .08 | .10 | .12 | .14 | .16 | .18 | .25 | .37 |
| Armour & Plastic | 3/8 | .06 | .07 | .08 | .09 | .12 | .16 | .20 | .25 | – | – | – |
| Sheath | 1/2 | .08 | .10 | .14 | – | – | – | – | – | – | – | – |
| Aluminum with | – | – | – | – | – | – | – | – | – | – | – | – |
| Plastic Sheath | 1/4 | – | – | .05 | .06 | .08 | .08 | .09 | .11 | .12 | .18 | .25 |
| Only | – | – | – | – | – | – | – | – | – | – | – | – |
| Aluminum with | – | – | – | – | – | – | – | – | – | – | – | – |
| Armour & Plastic | 1/4 | – | – | .06 | .07 | .09 | .09 | .10 | .13 | .14 | .20 | .28 |
| Sheath | – | – | – | – | – | – | – | – | – | – | – | – |
| Plastic Mylar | – | – | – | – | – | – | – | – | – | – | – | – |
| Envel. & Vinyl | 1/4 | .03 | .04 | .05 | .06 | .07 | .08 | .09 | .10 | .12 | .14 | .20 |
| Jacket | 3/8 | .04 | .04 | .06 | .07 | .09 | .09 | .11 | .12 | – | – | – |

| Fitting Type | Hours Each |
|---|---|
| Standard Indoor Junction Box 12″ x 18″ x 6″ | 2.5 |
| Weather Tight Junction Box 12″ x 18″ x 6″ | 4.0 |
| Union Box 10″ x 5″ x 5″ | 2.0 |
| Neoprene Grommets | 0.3 |
| Cast Aluminum Connector | 0.5 |
| Galvanized Connector and Neoprene Weatherproof Bushing | 0.9 |

Manhours include checking out of storage, handling to erection location, and complete installation with average time allowed for make-up.

Manhours do not include installation of instruments, cable tray or supports. See respective tables for these time frames.

# INSTALLATION OF SINGLE TUBING, FITTINGS, AND VALVES

MANHOURS PER UNITS LISTED

| Material Item | Unit | Size 1/4″ | 3/8″ | 1/2″ | 3/4″ | 1″ |
|---|---|---|---|---|---|---|
| Copper Tubing | LF. | .30 | .35 | .40 | .46 | .52 |
| Aluminum Tubing | LF. | .30 | .35 | .40 | .46 | .52 |
| Stainless Steel Tubing | LF. | .60 | .65 | .75 | .82 | .90 |
| Plastic Tubing | LF. | .30 | .35 | .40 | .46 | .52 |
| Steel Tubing | LF. | .40 | .48 | .53 | .59 | .65 |
| Fittings up to 600 Lb. | Ea. | .40 | .40 | .40 | .50 | .50 |
| Fittings over 600 Lb. | Ea. | .40 | .40 | .40 | .50 | .50 |
| Valves–Screwed | Ea. | 1.00 | 1.00 | 1.00 | 1.20 | 1.30 |
| Valves–Flanged | Ea. | 1.60 | 1.60 | 1.60 | 1.60 | 1.70 |

Manhours include checking out of storage, handling to erection location, and complete installation with average time allowed for make-up.

Manhours do not include installation of instruments, cable tray, or supports. See respective tables for these time frames.

# Section 6

# ELECTRICAL INSTALLATION

The electrical power and lighting portion of a project is generally let to a subcontractor who specializes in these type installations and is rarely installed by the mechanical contractor. However, the manhour tables in this section have been included to assist the mechanical estimator in preparation of an electrical estimate, for actual performance or as a check against an electrical subcontractor's estimate. These tables pertain to heating, ventilating, and air-conditioning systems in process or industrial plants, but they are not intended to suffice for the total installation of electrical power for such plants.

All labor for handling, hauling, setting and aligning, or installing has been given due consideration in the manhour tables in accordance with the notes that appear with each.

# CONDUIT, BOXES, & FITTINGS—OCTAGON, SQUARE, HANDY & SWITCH BOXES, & COVERS

MANHOURS PER UNITS LISTED

| Item Description | Unit | Heights To | | | |
|---|---|---|---|---|---|
| | | 10′ | 15′ | 20′ | 25′ |
| 4-Inch Octagon Boxes and Covers | | | | | |
| Octagon Box | Ea. | .32 | .33 | .34 | .35 |
| Octagon Box Extension | Ea. | .23 | .23 | .24 | .24 |
| Round Device Cover | Ea. | .22 | .23 | .24 | .25 |
| Round Blank Cover | Ea. | .15 | .15 | .16 | .17 |
| Round Cover with K.O. | Ea. | .15 | .15 | .16 | .17 |
| Round Switch or Receptical Cover | Ea. | .11 | .11 | .12 | .13 |
| Round Swivel Hanger Cover | Ea. | .11 | .11 | .12 | .13 |
| 4-Inch Square Boxes and Covers | | | | | |
| Square Box | Ea. | .29 | .30 | .32 | .33 |
| Square Box Extension | Ea. | .22 | .23 | .24 | .25 |
| Square Cover, 1-Device | Ea. | .22 | .23 | .24 | .25 |
| Square Cover, 2-Device | Ea. | .22 | .23 | .24 | .25 |
| Square offset Cover, 1-Device | Ea. | .44 | .46 | .48 | .50 |
| Square Blank Cover | Ea. | .15 | .15 | .16 | .17 |
| Square Cover with K.O. | Ea. | .15 | .15 | .16 | .17 |
| Square Swivel Hanger Cover | Ea. | .16 | .16 | .17 | .18 |
| 4-11/16-Inch Square Boxes and Covers | | | | | |
| Square Box | Ea. | .29 | .30 | .32 | .33 |
| Square Box Extension | Ea. | .22 | .23 | .24 | .25 |
| Square Cover, 1-Device | Ea. | .22 | .23 | .24 | .25 |
| Square Cover, 2-Device | Ea. | .22 | .23 | .24 | .25 |
| Square Blank Cover | Ea. | .15 | .15 | .16 | .17 |
| Square Blank Cover with K.O. | Ea. | .15 | .15 | .16 | .17 |
| Handy Boxes | | | | | |
| Handy Box | Ea. | .22 | .23 | .24 | .25 |
| Handy Box Extension | Ea. | .15 | .15 | .16 | .17 |
| Handy Box Cover | Ea. | .15 | .15 | .16 | .17 |
| Sectional Switch Box with K.O. | Ea. | .22 | .23 | .24 | .25 |
| Sectional Switch Box with Clamps | Ea. | .30 | .30 | .32 | .33 |

Manhours include checking out of job storage, handling, hauling, and installing items as outlined.

Manhours do not include installation of conduit, wire, or scaffolding. See respective tables for these time requirements.

# CONDUIT, BOXES, & FITTINGS—GANG BOXES & COVERS

MANHOURS PER UNIT LISTED

| Item Description | Unit | Heights To | | | |
|---|---|---|---|---|---|
| | | 10′ | 15′ | 20′ | 25′ |
| Gang Boxes | | | | | |
| 2-Gang | Ea. | .41 | .42 | .45 | .46 |
| 3-Gang | Ea. | .51 | .53 | .56 | .58 |
| 4-Gang | Ea. | .65 | .68 | .70 | .72 |
| 5-Gang | Ea. | .80 | .83 | .85 | .87 |
| 6-Gang | Ea. | .96 | 1.00 | 1.02 | 1.04 |
| 7-Gang | Ea. | 1.15 | 1.20 | 1.22 | 1.25 |
| 8-Gang | Ea. | 1.38 | 1.44 | 1.46 | 1.50 |
| Gang Device Covers | | | | | |
| 2-Gang Cover | Ea. | .15 | .15 | .16 | .17 |
| 3-Gang Cover | Ea. | .15 | .15 | .16 | .17 |
| 4-Gang Cover | Ea. | .15 | .15 | .16 | .17 |
| 5-Gang Cover | Ea. | .15 | .15 | .16 | .17 |
| 6-Gang Cover | Ea. | .23 | .23 | .24 | .25 |
| 7-Gang Cover | Ea. | .34 | .34 | .36 | .38 |
| 8-Gang Cover | Ea. | .51 | .51 | .54 | .57 |
| Gang Cover Surface Mounted Device | | | | | |
| 2-Surface Mounted Device | Ea. | .18 | .18 | .19 | .20 |
| 3-Surface Mounted Device | Ea. | .18 | .18 | .19 | .20 |
| 4-Surface Mounted Device | Ea. | .18 | .18 | .19 | .20 |
| 5-Surface Mounted Device | Ea. | .18 | .18 | .19 | .20 |
| 6-Surface Mounted Device | Ea. | .28 | .28 | .29 | .30 |
| Gang Blank Cover | | | | | |
| 2-Gang Blank Cover | Ea. | .15 | .15 | .16 | .17 |
| 3-Gang Blank Cover | Ea. | .15 | .15 | .16 | .17 |
| 4-Gang Blank Cover | Ea. | .15 | .15 | .16 | .17 |
| 5-Gang Blank Cover | Ea. | .15 | .15 | .16 | .17 |
| 6-Gang Blank Cover | Ea. | .23 | .23 | .24 | .25 |
| 7-Gang Blank Cover | Ea. | .34 | .34 | .36 | .38 |
| 8-Gang Blank Cover | Ea. | .51 | .51 | .54 | .57 |

Manhours include checking out of job storage, handling, hauling, and installing items as outlined.

Manhours do not include installation of conduit, wire, or scaffolding. See respective tables for these time frames.

# CONDUIT, BOXES, & FITTINGS—SHEET METAL BOXES FOR BRANCH ROUGH-IN

MANHOURS PER UNITS LISTED

| Item Description | Unit | Heights To | | | |
|---|---|---|---|---|---|
| | | 10′ | 15′ | 20′ | 25′ |
| S.C. Pull Boxes | | | | | |
| 4″ x 4″ x 4″ | Ea. | .68 | .71 | .73 | .75 |
| 4″ x 6″ x 4″ | Ea. | .68 | .71 | .73 | .75 |
| 6″ x 6″ x 4″ | Ea. | .68 | .71 | .73 | .75 |
| 6″ x 8″ x 4″ | Ea. | .72 | .76 | .78 | .80 |
| 8″ x 8″ x 4″ | Ea. | .72 | .76 | .78 | .80 |
| 8″ x 12″ x 4″ | Ea. | .90 | .95 | .98 | 1.01 |
| 12″ x 12″ x 4″ | Ea. | 1.08 | 1.13 | 1.16 | 1.19 |
| 12″ x 24″ x 4″ | Ea. | 1.71 | 1.80 | 1.85 | 1.91 |
| 12″ x 12″ x 6″ | Ea. | 1.35 | 1.42 | 1.46 | 1.50 |
| 12″ x 18″ x 6″ | Ea. | 1.53 | 1.61 | 1.66 | 1.71 |
| 12″ x 24″ x 6″ | Ea. | 1.71 | 1.80 | 1.85 | 1.91 |
| 18″ x 24″ x 6″ | Ea. | 2.25 | 2.36 | 2.43 | 2.50 |
| 18″ x 30″ x 6″ | Ea. | 2.48 | 2.60 | 2.68 | 2.76 |
| 24″ x 36″ x 6″ | Ea. | 3.38 | 3.55 | 3.66 | 3.77 |
| Hinge Cover Boxes | | | | | |
| 6″ x 6″ x 4″ | Ea. | .68 | .71 | .73 | .75 |
| 6″ x 8″ x 4″ | Ea. | .72 | .76 | .78 | .80 |
| 8″ x 8″ x 4″ | Ea. | .72 | .76 | .78 | .80 |
| 8″ x 12″ x 4″ | Ea. | .90 | .95 | .98 | 1.01 |
| 12″ x 12″ x 4″ | Ea. | 1.08 | 1.13 | 1.16 | 1.19 |
| 12″ x 18″ x 4″ | Ea. | 1.53 | 1.61 | 1.66 | 1.71 |
| 12″ x 24″ x 4″ | Ea. | 1.71 | 1.80 | 1.85 | 1.91 |
| 12″ x 12″ x 6″ | Ea. | 1.35 | 1.42 | 1.46 | 1.50 |
| 12″ x 18″ x 6″ | Ea. | 1.53 | 1.61 | 1.66 | 1.71 |
| 12″ x 24″ x 6″ | Ea. | 1.71 | 1.80 | 1.85 | 1.91 |
| 18″ x 24″ x 6″ | Ea. | 2.25 | 2.36 | 2.43 | 2.50 |

Manhours include checking out of job storage, handling, hauling, and installing items as outlined.

Manhours do not include installation of conduit, wire, or scaffolding. See respective tables for these time requirements.

# CONDUIT, BOXES, & FITTINGS—INSTALLING CONDUIT

MANHOURS PER HUNDRED LINEAR FEET

| Item Description | Size Inches | Heights To | | | |
|---|---|---|---|---|---|
| | | 10′ | 15′ | 20′ | 25′ |
| Rigid Galvanized–Conduit | 1/2 | 13.25 | 13.50 | 13.90 | 14.20 |
| Rigid Galvanized–Conduit | 3/4 | 16.30 | 16.65 | 17.15 | 17.50 |
| Rigid Galvanized–Conduit | 1 | 23.50 | 24.00 | 24.70 | 25.20 |
| Rigid Galvanized–Conduit | 1-1/4 | 24.25 | 24.75 | 25.50 | 26.00 |
| Rigid Galvanized–Conduit | 1-1/2 | 32.35 | 33.00 | 34.00 | 34.65 |
| Rigid Galvanized–Conduit | 2 | 37.50 | 38.25 | 39.40 | 40.15 |
| Rigid Galvanized–Conduit | 2-1/2 | 44.10 | 45.00 | 46.35 | 47.15 |
| Rigid Galvanized–Conduit | 3 | 58.80 | 60.00 | 61.80 | 63.00 |
| Rigid Galvanized–Conduit | 3-1/2 | 68.60 | 70.00 | 72.10 | 73.50 |
| Rigid Galvanized–Conduit | 4 | 73.50 | 75.00 | 77.25 | 78.75 |
| Rigid Galvanized–Conduit | 5 | 83.30 | 85.00 | 87.55 | 89.25 |
| Rigid Galvanized–Conduit | 6 | 88.20 | 90.00 | 92.70 | 94.50 |

Manhours include checking out of job storage, handling, hauling, and installing conduit as outlined.

For overhead work, add 20% to manhours.

In the case of parallel runs that are to be installed at the same time, apply the following percentages of manhours for additional runs:

For Second Parallel Run–95%
For Third Parallel Run–91%
For Fourth Parallel Run–87%
For Fifth Parallel Run–84%
For Each Additional Run Above Five–80%

For installation of steel conduit, use 107% of manhours.

Manhours do not include cutting, reaming, threading, and bending of conduit, or handling of fittings, and make-up of joints or connections. See respective tables for these manhours.

# CONDUIT BOXES, & FITTINGS—CUTTING, REAMING, & THREATING CONDUIT,& MAKING-ON OF JOINT

MANHOURS EACH

| Conduit Size Inches | Cut, Ream, and Thread | | | Make-On or Thread-On Heights To | | | |
|---|---|---|---|---|---|---|---|
| | Cutting Only | Ream & Thread | Combined Operation | 10′ | 15′ | 20′ | 25′ |
| 1/2 | .20 | .30 | .50 | .36 | .39 | .40 | .41 |
| 3/4 | .20 | .30 | .50 | .36 | .40 | .41 | .42 |
| 1 | .20 | .30 | .50 | .60 | .65 | .68 | .69 |
| 1-1/4 | .23 | .35 | .58 | .60 | .65 | .68 | .71 |
| 1-1/2 | .25 | .43 | .68 | .72 | .78 | .81 | .83 |
| 2 | .30 | .45 | .75 | .84 | .91 | .95 | .97 |
| 2-1/2 | .30 | .45 | .75 | .90 | .97 | 1.00 | 1.04 |
| 3 | .40 | .50 | .90 | 1.08 | 1.17 | 1.22 | 1.24 |
| 3-1/2 | .40 | .60 | 1.00 | 1.14 | 1.24 | 1.28 | 1.31 |
| 4 | .50 | .80 | 1.30 | 1.20 | 1.30 | 1.35 | 1.38 |
| 5 | .63 | 1.00 | 1.63 | 1.50 | 1.63 | 1.69 | 1.73 |
| 6 | .75 | 1.20 | 1.95 | 1.80 | 1.95 | 2.03 | 2.07 |

Cutting, reaming, and threading manhours include measuring, checking, cutting with hand saw, and threading and reaming by hand. If cutting is accomplished with power band saw and threading and reaming with power machine, use 60% of manhours.

Make-on or thread-on manhours include checking fitting out of barge storage, rigging, picking, and installing, applying joint sealer to threads, and single hub make-on. Manhours are average for one thread-on only and for any type fitting. It must be remembered that a coupling, a 90° Ell, or a 45° Ell has two thread-ons, a cap has one thread-on, etc.

# CONDUIT BENDING

MANHOURS PER BEND

| Conduit Size Inches | Number and Type of bends | | | | |
|---|---|---|---|---|---|
| | 1, 2, 3, & 4 | 5 | 6 | 7 | 8 |
| 1/2 | .30 | .68 | .41 | .40 | .68 |
| 3/4 | .35 | .77 | .54 | .45 | .77 |
| 1 | .55 | 1.11 | .78 | .60 | 1.11 |
| 1-1/4 | .62 | 2.21 | 1.55 | .70 | 2.21 |
| 1-1/2 | .80 | 2.47 | 1.73 | .90 | 2.47 |
| 2 | 1.00 | 2.89 | 2.00 | 1.20 | 2.89 |
| 2-1/2 | 1.25 | 3.32 | 2.23 | 1.60 | 3.32 |
| 3 | 1.50 | 3.74 | 2.62 | 1.85 | 3.74 |
| 3-1/2 | 1.80 | 4.42 | 3.10 | 2.00 | 4.42 |
| 4 | 2.50 | 4.85 | 3.40 | 2.20 | 4.85 |
| 5 | 3.13 | – | – | – | – |
| 6 | 3.75 | – | – | – | – |

Manhours include use of standard lengths or pieces of conduit in proximity of final installation. Bends are made by hand as single or multiple operations on standard portable equipment. Manhours for 1/2-inch through 1-inch bends are for general use hickie.

Manhours do not include cutting, reaming, threading, make-ons, or installation. See respective tables for these time frames.

# STANDARD TYPES OF BENDS

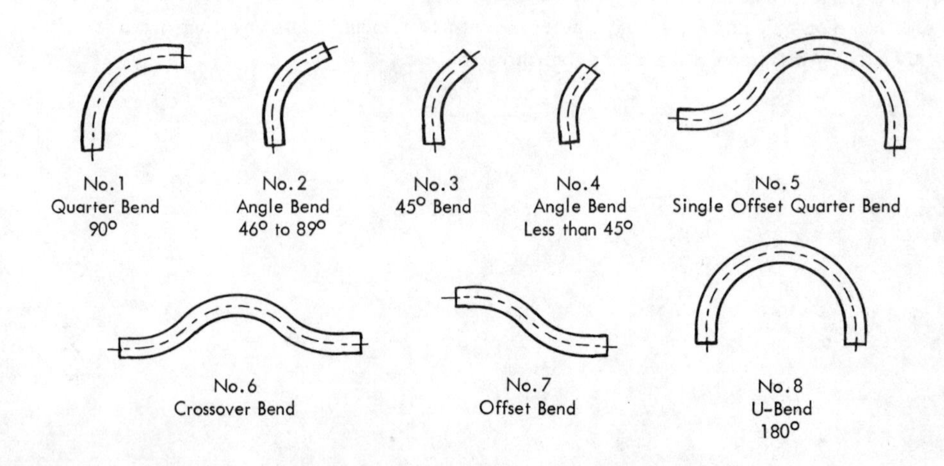

# SERVICE & FEEDER WIRING
# WIRE PULLING—SIMPLE LAY-OUT

**Stranded or Solid Wire**
**Rubber or Thermoplastic Insulated Covered**

MANHOURS FOR NUMBER OF WIRES AND RUNS

| Wire Size | 50′ Conduit Runs | | | 100′ Conduit Runs | | | 200′ Conduit Runs | | |
|---|---|---|---|---|---|---|---|---|---|
| | 2-Wire | 3-Wire | 4-Wire | 2-Wire | 3-Wire | 4-Wire | 2-Wire | 3-Wire | 4-Wire |
| #18 | .54 | .72 | .99 | 1.04 | 1.50 | 1.89 | 2.02 | 2.87 | 3.67 |
| #16 | .66 | .88 | 1.21 | 1.27 | 1.83 | .231 | 2.47 | 3.50 | 4.49 |
| #14 | .81 | 1.08 | 1.48 | 1.54 | 2.22 | 2.80 | 3.00 | 4.26 | 5.44 |
| #12 | .99 | 1.57 | 1.80 | 1.98 | 2.70 | 3.40 | 3.64 | 5.22 | 6.64 |
| #10 | 1.39 | 1.98 | 2.52 | 2.64 | 3.78 | 4.85 | 5.12 | 7.38 | 9.28 |
| #8 | 1.68 | 2.40 | 3.04 | 3.20 | 4.56 | 5.80 | 6.20 | 8.88 | 11.28 |
| #6 | 1.74 | 2.49 | 3.16 | 3.30 | 4.74 | 6.00 | 6.40 | 9.18 | 11.68 |
| #4 | 2.09 | 2.97 | 3.76 | 3.98 | 5.67 | 7.20 | 7.72 | 10.98 | 13.92 |
| #2 | 2.32 | 3.30 | 4.16 | 4.42 | 6.27 | 8.00 | 8.56 | 12.24 | 15.44 |
| #1 | 2.66 | 3.78 | 4.80 | 5.06 | 7.23 | 9.20 | 9.84 | 14.10 | 17.84 |
| #1/0 | 3.37 | 4.80 | 6.08 | 6.42 | 9.15 | 11.60 | 12.44 | 17.76 | 22.48 |
| #2/0 | 3.70 | 5.28 | 6.68 | 7.06 | 10.05 | 12.76 | 13.68 | 19.56 | 24.72 |
| #3/0 | 4.40 | 6.27 | 7.96 | 8.38 | 11.94 | 15.16 | 16.28 | 23.22 | 29.44 |
| #4/0 | 4.98 | 7.11 | 9.00 | 9.48 | 13.53 | 17.16 | 18.47 | 26.28 | 33.36 |
| 250 MCM | 5.21 | 7.44 | 9.40 | 9.92 | 14.13 | 17.96 | 19.68 | 28.04 | 34.80 |
| 300 MCM | 5.57 | 7.96 | 10.06 | 10.61 | 15.12 | 19.22 | 21.06 | 30.00 | 37.24 |
| 350 MCM | 5.79 | 8.28 | 10.48 | 11.04 | 15.72 | 19.96 | 21.40 | 30.60 | 38.72 |
| 400 MCM | 6.20 | 8.86 | 11.21 | 11.81 | 16.82 | 21.36 | 22.90 | 32.74 | 41.43 |
| 500 MCM | 6.72 | 9.60 | 12.16 | 12.80 | 18.24 | 23.16 | 24.84 | 35.46 | 44.88 |
| 600 MCM | 7.19 | 10.26 | 13.00 | 13.68 | 19.50 | 24.76 | 26.56 | 37.86 | 48.00 |
| 750 MCM | 9.26 | 13.23 | 16.76 | 17.64 | 25.14 | 31.92 | 34.20 | 48.90 | 61.92 |
| 1000 MCM | 11.12 | 15.87 | 20.08 | 21.16 | 30.15 | 38.32 | 41.08 | 58.68 | 74.32 |

A simple lay-out is one that presents normal working conditions and conduit runs free from obstacles such as fittings or pull boxes at right angle turns.

Manhours include checking out of job storage, handling, hauling, and installing wire as outlined.

If duplicate pulls are to be made, deduct the following percentages from the manhours to compensate for equipment set-up time:

#18 to #8—4%
#6 to #4—5%
#2 to #1—7%
#1/0 to 1000 MCM—9%

# SERVICE & FEEDER WIRING
# WIRE PULLING—COMPLEX LAY-OUT
### Stranded or Solid Wire
### Rubber or Thermoplastic Insulated Covered

MANHOURS FOR NUMBER OF WIRES AND RUNS

| Wire | 50′ Conduit Runs | | | 100′ Conduit Runs | | | 200′ Conduit Runs | | |
|---|---|---|---|---|---|---|---|---|---|
| Size | 2-Wire | 3-Wire | 4-Wire | 2-Wire | 3-Wire | 4-Wire | 2-Wire | 3-Wire | 4-Wire |
| #18 | .78 | 1.11 | 1.41 | 1.48 | 2.12 | 2.69 | 2.88 | 4.12 | 5.23 |
| #16 | .95 | 1.35 | 1.71 | 1.80 | 2.57 | 3.26 | 3.49 | 4.99 | 6.34 |
| #14 | 1.16 | 1.65 | 2.10 | 2.20 | 3.15 | 4.00 | 4.28 | 6.12 | 7.76 |
| #12 | 1.41 | 2.01 | 2.56 | 2.63 | 3.84 | 4.80 | 5.20 | 7.44 | 9.44 |
| #10 | 1.98 | 2.82 | 3.60 | 3.78 | 5.40 | 6.84 | 6.52 | 10.50 | 13.28 |
| #8 | 2.40 | 3.42 | 4.34 | 4.58 | 6.54 | 8.28 | 8.83 | 12.61 | 16.08 |
| #6 | 2.48 | 3.54 | 4.50 | 4.72 | 6.75 | 9.56 | 9.66 | 13.08 | 16.64 |
| #4 | 2.98 | 4.22 | 5.40 | 5.68 | 8.10 | 10.28 | 11.00 | 15.72 | 19.92 |
| #2 | 3.31 | 4.71 | 5.98 | 6.30 | 9.00 | 11.40 | 12.24 | 17.96 | 22.08 |
| #1 | 3.80 | 5.43 | 6.88 | 7.24 | 10.35 | 13.12 | 14.04 | 20.10 | 25.44 |
| #1/0 | 4.81 | 6.87 | 9.70 | 9.16 | 13.08 | 16.56 | 17.76 | 25.38 | 32.16 |
| #2/0 | 5.29 | 7.56 | 9.58 | 10.08 | 14.40 | 18.24 | 19.56 | 27.96 | 35.36 |
| #3/0 | 6.29 | 8.97 | 11.38 | 11.98 | 17.10 | 21.68 | 23.24 | 33.18 | 42.08 |
| #4/0 | 7.11 | 10.14 | 12.88 | 13.54 | 19.35 | 24.52 | 26.28 | 37.56 | 47.60 |
| 250 MCM | 7.49 | 10.62 | 13.46 | 14.18 | 20.25 | 25.64 | 27.52 | 39.30 | 49.76 |
| 300 MCM | 7.96 | 11.36 | 14.40 | 15.17 | 21.67 | 27.43 | 29.45 | 42.05 | 53.24 |
| 350 MCM | 8.27 | 11.82 | 14.98 | 15.76 | 22.50 | 28.52 | 30.56 | 43.68 | 55.36 |
| 400 MCM | 8.85 | 12.65 | 16.03 | 16.86 | 24.08 | 30.52 | 32.70 | 46.74 | 59.24 |
| 500 MCM | 9.60 | 12.71 | 17.36 | 18.28 | 26.10 | 33.08 | 35.48 | 50.64 | 64.16 |
| 600 MCM | 10.26 | 14.64 | 18.56 | 19.54 | 27.90 | 35.36 | 37.92 | 54.12 | 68.56 |
| 750 MCM | 13.23 | 18.90 | 23.94 | 25.20 | 36.00 | 45.60 | 48.88 | 69.84 | 88.48 |
| 1000 MCM | 15.88 | 22.68 | 28.72 | 30.24 | 43.20 | 54.72 | 58.68 | 83.82 | 106.16 |

A complex lay-out is one that requires pulling cable for various locations under adverse working conditions or through pull boxes at right angles. There are many and varying degrees of complexity to be considered.

Manhours include checking out of job storage, handling, hauling, and installing wire as outlined.

If duplicate pulls are to be made, deduct the following percentages from the manhours to compensate for equipment setup time:

#18 to #8–4%

#6 to #4–5%

#2 to #1–7%

#1/0 to 1000 MCM–9%

# CABLE INSTALLATION
# FLEXIBLE METALLIC ARMORED CABLE
# (BX AND BXL)

MANHOURS REQUIRED FOR NUMBER OF CONDUCTORS AND RUNS

| Number & Size Conductors | 50′ Runs | | 100′ Runs | | 200′ Runs | |
|---|---|---|---|---|---|---|
| | Concealed | Exposed | Concealed | Exposed | Concealed | Exposed |
| **BX Cable–Copper Conductor** | | | | | | |
| 2/#14 | 2.75 | 2.36 | 5.25 | 4.50 | 9.98 | 8.56 |
| 3/#14 | 2.99 | 2.60 | 5.70 | 4.95 | 10.84 | 9.40 |
| 4/#14 | | | | | | |
| 2/#12 | 2.99 | 2.60 | 5.70 | 4.95 | 10.84 | 9.40 |
| 3/#12 | 3.54 | 2.75 | 6.75 | 5.25 | 12.82 | 9.98 |
| 4/#12 | | | | | | |
| 2/#10 | 3.94 | 3.94 | 7.50 | 7.50 | 14.26 | 14.26 |
| 3/#10 | 4.72 | 4.41 | 9.00 | 8.40 | 17.10 | 15.96 |
| 4/#10 | | | | | | |
| 2/#8 | 5.12 | 4.92 | 9.75 | 9.38 | 18.52 | 17.82 |
| 3/#8 | 5.90 | 5.51 | 11.25 | 10.50 | 21.38 | 19.96 |
| 4/#8 | | | | | | |
| **BXL Cable–Copper Conductor** | | | | | | |
| 2/#14 | 2.99 | 2.60 | 5.70 | 4.95 | 10.84 | 9.40 |
| 3/#14 | 3.15 | 2.75 | 6.00 | 5.25 | 11.40 | 9.98 |
| 2/#12 | 3.15 | 2.75 | 6.00 | 5.25 | 11.40 | 9.98 |
| 3/#12 | 3.74 | 3.35 | 7.13 | 6.38 | 13.54 | 12.12 |
| 2/#10 | 4.33 | 3.94 | 8.25 | 7.50 | 15.68 | 14.26 |
| 3/#10 | 5.12 | 4.72 | 9.75 | 9.00 | 18.52 | 17.10 |
| 2/#8 | 5.32 | 5.12 | 10.13 | 9.75 | 19.24 | 18.52 |
| 3/#8 | 6.30 | 5.90 | 12.00 | 11.25 | 22.80 | 21.38 |
| **Armored Single Wire–Copper Conductor** | | | | | | |
| 1/#8 | 2.00 | 1.70 | 3.82 | 3.44 | 8.48 | 7.63 |
| 1/#6 | 2.50 | 2.13 | 4.78 | 4.30 | 10.61 | 9.55 |
| 1/#4 | 3.00 | 3.53 | 5.73 | 5.16 | 12.72 | 11.45 |

Manhours include checking out of job storage, handling, hauling, and installing of items as outlined.

Manhours are average for length of runs as shown in the table regardless of height.

Manhours do not include splicing or connecting to switches or receptacles. See respective tables for these time requirements.

# CABLE CONNECTORS & LUGS

MANHOURS REQUIRED EACH

| Wire Size | Type or Connector or Lugs | | | |
|---|---|---|---|---|
| | 1 | 2 | 3 | 4 |
| #6 | .54 | .45 | .27 | .23 |
| #4 | .63 | .54 | .27 | .23 |
| #2 | .72 | .63 | .27 | .23 |
| #1 | 1.04 | .90 | .54 | .45 |
| #1/0 | 1.13 | 1.04 | .68 | .59 |
| #2/0 | 1.35 | 1.13 | .81 | .68 |
| #3/0 | 1.58 | 1.35 | .99 | .81 |
| #4/0 | 1.80 | 1.58 | 1.13 | .90 |
| 250 MCM | 2.25 | 1.80 | 1.40 | 1.13 |
| 300 MCM | 2.48 | 2.03 | 1.58 | 1.35 |
| 350 MCM | 2.70 | 2.48 | 1.71 | 1.44 |
| 400 MCM | 3.15 | 2.70 | 1.80 | 1.58 |
| 500 MCM | 3.60 | 3.15 | 2.07 | 1.80 |
| 750 MCM | 4.50 | 3.60 | 3.15 | 2.48 |
| 1000 MCM | 5.40 | 4.50 | 3.60 | 2.93 |

1. V-Bolt Connector—Tape Wrapped
2. Two-Way Compression Connector—Tape Wrapped
3. Bolt Lugs
4. Compression Lugs

# CABLE VERTICAL RISER SUPPORTS

MANHOURS REQUIRED EACH

| Size Inches | Height To | | | |
|---|---|---|---|---|
| | 10′ | 15′ | 20′ | 25′ |
| 1-1/4 | .77 | .79 | .81 | .83 |
| 1-1/2 | .99 | 1.02 | 1.05 | 1.08 |
| 2 | 1.49 | 1.53 | 1.58 | 1.63 |
| 2-1/2 | 1.98 | 2.04 | 2.10 | 2.16 |
| 3 | 2.48 | 2.55 | 2.63 | 2.71 |
| 3-1/2 | 2.97 | 3.06 | 3.15 | 3.24 |
| 4 | 3.83 | 3.95 | 4.07 | 4.19 |
| 5 | 4.73 | 4.87 | 5.02 | 5.18 |
| 6 | 5.85 | 6.03 | 6.21 | 6.40 |

Manhours include checking out of job storage, handling, hauling, and installing of items listed. Manhours do not include cable installation. See respective table for this time requirement.

# SWITCHES & PLATES

MANHOURS REQUIRED EACH

| Type | Item Description | Device Only | Standard Plates Only | Complete With Plates |
|---|---|---|---|---|
| 10A | 1-Pole Switch | .22 | .15 | .37 |
| 10A | 2-Pole Switch | .75 | .15 | .90 |
| 10A | 3-Way Switch | .53 | .15 | .68 |
| 10A | 4-Way Switch | .75 | .15 | .90 |
| 10A | 2-Pole, 3-Point Switch | .60 | .15 | .75 |
| 10A | 2-Pole, 4-Point Switch | .60 | .15 | .75 |
| 20A | 1-Pole Switch | .45 | .15 | .60 |
| 20A | 2-Pole Switch | 1.05 | .15 | 1.20 |
| 20A | 3-Way Switch | .68 | .15 | .83 |
| 20A | 4-Way Switch | 1.05 | .15 | 1.20 |
| 30A | 1-Pole Switch | .60 | .15 | .75 |
| 30A | 2-Pole Switch | 1.13 | .15 | 1.28 |
| 30A | 3-Way Switch | .90 | .15 | 1.05 |
| 30A | 4-Way Switch | 1.13 | .15 | 1.28 |
| 10A | 1-Pole, Ceiling Pull Switch | – | – | .75 |
| 10A | 3-Way, Ceiling Pull Switch | – | – | 1.13 |
| – | Pendant Switch and Cord | – | – | .75 |
| – | Pendant Switch and Heavy Duty Cord | – | – | 1.05 |
| – | Canopy Switch (Assembled with Fixture) | – | – | .23 |
| – | Door Switch and Box | – | – | 2.25 |
| 10A | Mark Time Switch | – | – | .68 |
| 20A | Mark Time Switch | – | – | .90 |

Manhours include checking out of job storage, handling, hauling, and installing of items as outlined.

*Device Only* includes installation of switch and making up.

*Standard Plates Only* includes a separate operation for installing plate.

*Complete With Plates* includes manhours for the installation of device and plate as one operation.

# RECEPTACLES, PLATES, MISCELLANEOUS OUTLETS, & COVERS

MANHOURS REQUIRED EACH

| Type | Item Description | Device Only | Standard Plate Only | Complete With Plates |
|---|---|---|---|---|
| 10A | Duplex Receptacle | .27 | .15 | .42 |
| 20A | 2-Wire Receptacle | .42 | .15 | .57 |
| 30A | 2-Wire Receptacle | .68 | .15 | .83 |
| 10A | 3-Wire Receptacle | .45 | .15 | .60 |
| 20A | 3-Wire Receptacle | .57 | .15 | .72 |
| 30A | 3-Wire Receptacle | .75 | .15 | .90 |
| 20A | 3-Wire Ground Type Receptacle | .53 | .15 | .68 |
| 30A | 3-Wire Ground Type Receptacle | .68 | .15 | .83 |
| 50A | 3-Wire Ground Type Receptacle | 1.05 | .15 | 1.20 |
| 20A | 4-Wire Ground Type Receptacle | .68 | .15 | .83 |
| 30A | 4-Wire Ground Type Receptacle | .83 | .15 | .98 |
| 50A | 4-Wire Ground Type Receptacle | 1.20 | .15 | 1.35 |
| 10A | Weatherproof Receptacle | – | – | .38 |
| 10A | Clock Hanger Outlet | .53 | .23 | .76 |
| 10A | Fan Hanger Outlet | .83 | .23 | 1.06 |
| – | Pilot Light Receptacle | .60 | .23 | .83 |
| – | Cover Socket–Ceiling | – | – | .60 |
| – | Floor Outlet with Plate–One Gang | – | – | 2.25 |
| – | Floor Outlet W.P.–Gang Type (Per Gang) | – | – | 1.88 |
| 15A | Receptacle in Floor Box | .68 | – | – |
| – | Receptacle Stand–Floor Box | .60 | – | – |
| – | Bell Nozzle–Floor Box | .53 | – | – |
| – | Switch and Receptacle Plate (Special) | – | .30 | – |
| – | Radio Outlet | – | – | .83 |

Manhours include checking out of job storage, handling, hauling, and installing of items as outlined.

*Device Only* includes installation of receptacle or outlet and make-up operation.

*Standard Plates Only* includes a separate operation for installing plate only.

*Complete With Plates* includes manhours for the installation of receptacles and plates as one operation.

# STANDARD PANELS & CABINETS

MANHOURS FOR PANELBOARD AND CABINET

| Number of 30-Amp 2-Wire Branch Circuits | Hours Required | |
|---|---|---|
| | Flush Mounted | Surface Mounted |
| 4 | 4.95 | 4.50 |
| 6 | 7.50 | 6.75 |
| 8 | 9.00 | 8.10 |
| 10 | 10.50 | 9.45 |
| 12 | 13.50 | 12.15 |
| 14 | 15.00 | 14.10 |
| 16 | 16.50 | 14.86 |
| 18 | 18.00 | 16.20 |
| 20 | 19.50 | 17.55 |
| 22 | 21.00 | 18.90 |
| 24 | 22.50 | 20.25 |
| 26 | 24.00 | 21.60 |
| 28 | 25.50 | 22.95 |
| 30 | 27.00 | 24.30 |
| 32 | 28.50 | 25.65 |
| 34 | 33.00 | 29.70 |
| 36 | 34.50 | 31.05 |
| 38 | 36.00 | 32.40 |
| 40 | 37.90 | 34.10 |
| 42 | 39.79 | 35.81 |

Manhours are for the installation of standard panelboards and cabinets having fuses only or fuses and switches in the branches, and having mains with lugs only, main fuses, main switches, or main switches, and fuses or circuit breaker.

Time has been allowed for the removal of insides and placing in temporary job storage, rigging, picking, and setting box in position, punching necessary knock-outs, locating and installing box, returning to storage, picking up insides, installing in cabinet, installing main pipe terminals, and installing cover plate.

Manhours do not include installation of fasteners, supports, sub-feeder terminals or sub-feeder pipe entrances. See respective tables for these time frames.

# HANGERS & FASTENERS

MANHOURS PER HUNDRED

| Size Inches | One Hole Strap Type | | | | Split Pipe Rings & Sockets | | | | Pipe Riser Clamps | | | |
|---|---|---|---|---|---|---|---|---|---|---|---|---|
| | Height To | | | | Height To | | | | Height To | | | |
| | 10′ | 15′ | 20′ | 25′ | 10′ | 15′ | 20′ | 25′ | 10′ | 15′ | 20′ | 25′ |
| 3/8 | 1.37 | 1.40 | 1.44 | 1.47 | – | – | – | – | – | – | – | – |
| 1/2 | 1.37 | 1.40 | 1.44 | 1.47 | – | – | – | – | – | – | – | – |
| 3/4 | 1.37 | 1.40 | 1.44 | 1.47 | – | – | – | – | – | – | – | – |
| 1 | 1.86 | 1.90 | 1.95 | 2.00 | – | – | – | – | – | – | – | – |
| 1-1/4 | 2.74 | 2.80 | 2.88 | 2.94 | 17.86 | 18.00 | 18.55 | 18.90 | 48.01 | 49.50 | 51.00 | 51.98 |
| 1-1/2 | 2.74 | 2.80 | 2.88 | 2.94 | 26.19 | 27.00 | 27.80 | 28.35 | 53.84 | 55.50 | 57.15 | 58.28 |
| 2 | 4.12 | 4.20 | 4.33 | 4.41 | 33.46 | 34.50 | 35.55 | 36.20 | 58.20 | 60.00 | 61.80 | 63.00 |
| 2-1/2 | 4.12 | 4.20 | 4.33 | 4.41 | 40.74 | 42.00 | 43.25 | 44.10 | 62.57 | 64.50 | 66.45 | 67.72 |
| 3 | 6.86 | 7.00 | 7.21 | 7.35 | 48.00 | 49.50 | 51.00 | 51.98 | 63.39 | 70.50 | 72.60 | 74.00 |
| 3-1/2 | 8.43 | 8.60 | 8.86 | 9.03 | 62.56 | 64.50 | 68.45 | 67.70 | 77.12 | 79.50 | 81.90 | 83.47 |
| 4 | 9.80 | 10.00 | 10.30 | 10.50 | 69.85 | 72.00 | 74.15 | 75.60 | 87.30 | 90.00 | 92.70 | 94.50 |

| Size Inches | Item Description | Height To | | | |
|---|---|---|---|---|---|
| | | 10′ | 15′ | 20′ | 25′ |
| – | Beam Clamps | 26.30 | 27.00 | 27.80 | 28.35 |
| 1/4 | Rod-Size Expansion Anchors | 29.30 | 30.00 | 30.90 | 31.50 |
| 3/8 | Rod-Size Expansion Anchors | 47.50 | 48.00 | 49.45 | 50.40 |
| 1/2 | Rod-Size Expansion Anchors | 56.50 | 57.00 | 58.70 | 59.85 |
| – | Concrete Inserts 3/8″ or 1/2″ Nuts | 36.75 | 37.50 | 38.65 | 39.40 |
| – | Ceiling Flanges and Sockets | 29.30 | 30.00 | 30.90 | 31.50 |

Manhours include checking out of job storage, handling, hauling, and installing items as outlined.

Manhours do not include installation of electrical devices. See respective tables for these time frames.

# MISCELLANEOUS FASTENERS

MANHOURS PER ITEM OR OPERATION

| Bolt Size Inches | Lead Expansion Anchors | Toggle Bolts | Wooden and Lag Screws | Mach. Screws In steel Drill & Tap | Mach. Bolt In Steel Av. 3/8″ Thick |
|---|---|---|---|---|---|
| 1/8″ | – | .11 | – | .32 | .24 |
| 3/16″ | .14 | .12 | – | .39 | .27 |
| 1/4″ | .15 | .14 | – | .42 | .30 |
| 5/16″ | .18 | – | – | .48 | .34 |
| 3/8″ | .25 | .18 | – | .58 | .38 |
| 7/16″ | .28 | – | – | .65 | .44 |
| 1/2″ | .28 | – | – | – | – |
| 5/8″ | .38 | – | – | – | – |
| #10 x 1″ | – | – | .03 | – | – |
| #12 x 1-1/4″ | – | – | .03 | – | – |
| 1/4″ x 1-1/2″ | – | – | .05 | – | – |
| 3/8″ x 2″ | – | – | .08 | – | – |
| 1/2″ x 2-1/2″ | – | – | .11 | – | – |

Manhours include checking out of job storage, handling, hauling, fabricating hole with power tool when required, and erection of anchor bolt or screw.

Manhours are average for heights to 25 feet.

Manhours do not include straps or installation of other supports. See respective tables for these time requirements.

# MOTOR STARTING SWITCHES & DIAL TYPE SPEED REGULATING RHEOSTATS

MANHOURS REQUIRED EACH

| Motor Starting Switches For 30-Amp AC Motors | | | Dial Type Speed Regulating Rheostats for 220-Volt, 3-Phase, Slip-Ring AC Induction Motors | |
|---|---|---|---|---|
| Type Switch & Motor | Soldered | Solderless | Motor HP | Mounting |
| 2-Pole, Single Phase | 5.63 | 5.10 | 1 | 6.60 |
| 3-Pole, 3-Phase or | | | 2-3 | 7.50 |
| 3-Wire, 2-Phase | 6.38 | 5.78 | 5 | 8.40 |
| 4-Pole, 4-Wire | | | 7-1/2 - 10 | 10.80 |
| 2-Phase | 6.75 | 6.08 | 15 | 12.60 |

## 30-AMP AC MAGNETIC SWITCHES

## 3-POLE, 220-VOLT AC MAGNETIC SWITCH

MANHOURS REQUIRED EACH

| Number of Poles | Mounting |
|---|---|
| 3 | 7.50 |
| 4 | 8.70 |
| – | – |
| – | – |

| Capacity of Switch Amperes | Mounting |
|---|---|
| 15 | 6.00 |
| 75 | 9.45 |
| 150 | 12.00 |
| 300 | 15.00 |

*Motor Starting Switches for 30-Amp, AC Motors* manhours include making connection at motor, and testing for direction of rotation for 30-amp, motor-starting switches.

*Dial Type Regulation Rheostats* manhours include mounting switch and rheostat and making connection at motor.

*30-Amp, AC Magnetic Switches* manhours include mounting and connecting with thermal cutouts or relays used as starters for small squirrel-cage motors, mounting and connecting push-button control station, and making connection at the motor and testing.

*3-Pole, 220-Volt AC Magnetic Switches* manhours include mounting and connecting push-button control station.

Manhours do not include installation of conduit, pulling of wire, or mounting of motors. See other tables for these time frames.

# STARTING COMPENSATORS FOR 3-PHASE, SQUIRREL-CAGE AC INDUCTION MOTORS

| Voltage | Motor Horsepower | Mounting |
|---|---|---|
| 220 | 7-1/2 - 10 | 8.25 |
| 220 | 15 | 10.50 |
| 220 | 20 - 25 | 11.70 |
| 220 | 30 | 12.75 |
| 220 | 40 | 14.25 |
| 220 | 50 | 15.00 |
| 220 | 75 | 16.50 |
| 220 | 100 | 21.00 |
| 440 | 7-1/2 - 10 - 15 | 8.25 |
| 440 | 20 - 25 | 10.50 |
| 440 | 30 | 11.25 |
| 440 | 40 - 50 | 12.00 |
| 440 | 75 | 14.25 |
| 440 | 100 | 15.75 |

Manhours include checking out of job storage, handling, hauling, mounting compensator, making connections at compensator and motor, and testing motor for direction of rotation.

For 2-phase, 4-wire motors, add 20% to manhours.

Manhours do not include installation of conduit, pulling of wire, or mounting or motors. See respective tables for these time frames.

# DC MOTOR RHEOSTATS & SWITCHES

MANHOURS REQUIRED EACH

| STARTING RHEOSTATS AND EXTERNALLY OPERATED SWITCHES | | | | | |
|---|---|---|---|---|---|
| Voltage | Motor Horsepower | Mounting | Voltage | Motor Horsepower | Mounting |
| 115 | 1/2 - 3/4 - 1 | 6.90 | 230 | 1/2 - 3/4 - 1 | 6.90 |
| 115 | 1-1/2 - 2 - 3 | 7.50 | 230 | 1-1/2 - 2 - 3 - 4 - 5 | 7.50 |
| 115 | 5 | 10.05 | 230 | 7-1/2 | 10.05 |
| 115 | 7-1/2 | 13.05 | 230 | 10 | 11.10 |
| 115 | 10 | 13.80 | 230 | 15 | 13.05 |
| 115 | 15 | 16.05 | 230 | 20 | 14.10 |
| 115 | 20 | 17.05 | 230 | 25 | 16.05 |
| 115 | 25 | 21.45 | 230 | 30 - 35 | 17.40 |
| – | – | – | 230 | 40 | 18.30 |
| – | – | – | 230 | 50 | 20.70 |

| Push-button Controlled Magnetic Switches and Line Switches | | | Speed Regulating Rheostats (Controlled by Resistance in Armature Circuit) and Externally Operated Switches | | |
|---|---|---|---|---|---|
| Voltage | Motor Horsepower | Mounting | Voltage | Motor Horsepower | Mounting |
| 115 | 1 - 2 - 3 | 12.00 | 115 | 1/2 - 3/4 - 1 | 7.95 |
| 115 | 5 | 14.10 | 115 | 1-1/2 - 2 - 3 | 9.00 |
| 115 | 7-1/2 - 10 | 16.50 | 115 | 5 | 12.00 |
| 230 | 1 - 2 - 3 - 5 | 12.00 | 115 | 7-1/2 - 10 | 15.00 |
| 230 | 7-1/2 - 10 | 14.10 | 230 | 1/2 - 3/4 - 1 | 7.95 |
| 230 | 15 | 16.05 | 230 | 1-1/2 - 2 - 3 | 9.00 |
| – | – | – | 230 | 5 | 9.60 |
| – | – | – | 230 | 7-1/2 - 10 | 12.00 |

*Starting Rheostats and Externally Operated Switches* manhours include mounting and connecting at switches, rheostats, and motors. For speed regulator with shunt field weakening instead of plain starting rheostats, add one (1) hour in each case.

*Push-Button Controlled Magnetic Switches and Line Switches* manhours include mounting and connecting push-button control station and connections at motor.

*Speed Regulating Rheostats and Externally Operated Switches* manhours include mounting and making connections at switch, rheostat and motor.

Manhours do not include conduit installation, pulling of wire, or mounting of motor. See other tables for these time frames.

# MOUNTING MOTORS—AC, 60-CYCLE, 2- AND 3-PHASE, 220, 440, OR 550 VOLTS

MANHOURS REQUIRED EACH

| Horse-power | RPM Rating and Mounting Hours | | | | | | |
|---|---|---|---|---|---|---|---|
| | 1750 | 1160 | 875 | 700 | 575 | 490 | 420 |
| 1 | 2.91 | 2.91 | 3.64 | 4.37 | 4.37 | 5.09 | 6.55 |
| 1-1/2 | 3.28 | 3.28 | 4.37 | 4.37 | 5.09 | 6.55 | 8.00 |
| 2 | 3.64 | 3.64 | 4.37 | 5.09 | 6.55 | 8.00 | 10.19 |
| 3 | 3.64 | 4.37 | 5.09 | 6.55 | 8.00 | 10.19 | 14.55 |
| 5 | 4.37 | 5.09 | 6.55 | 8.00 | 10.19 | 14.55 | 15.28 |
| 7-1/2 | 5.09 | 6.55 | 8.00 | 10.19 | 14.55 | 15.28 | 18.92 |
| 10 | 6.55 | 8.00 | 10.19 | 14.55 | 15.28 | 18.92 | 20.37 |
| 15 | 8.00 | 10.19 | 14.55 | 15.28 | 18.92 | 20.37 | 21.83 |
| 20 | 10.19 | 11.64 | 15.28 | 18.92 | 20.37 | 21.83 | 22.55 |
| 25 | 11.64 | 15.28 | 18.92 | 20.37 | 21.83 | 22.55 | 37.83 |
| 30 | 14.55 | 18.92 | 20.37 | 21.83 | 22.55 | 37.83 | 37.83 |
| 35 | 15.28 | 20.37 | 21.83 | 22.55 | 33.47 | 37.83 | 43.65 |
| 40 | 18.92 | 20.37 | 21.83 | 22.55 | 37.83 | 37.83 | 43.65 |
| 50 | 21.83 | 24.73 | 24.73 | 33.47 | 37.83 | 43.65 | 43.65 |
| 60 | 24.73 | 29.10 | 33.47 | 37.83 | 43.65 | 43.65 | 55.29 |
| 75 | 29.10 | 34.92 | 37.83 | 43.65 | 55.29 | 55.29 | 73.14 |
| 100 | 34.92 | 37.83 | 43.65 | 55.29 | 73.14 | 73.14 | 80.70 |
| 125 | 34.92 | 42.20 | 55.29 | 73.14 | 73.14 | 80.70 | 89.53 |
| 440- or 550-Volt Motors | | | | | | | |
| 150 | 37.83 | 42.20 | 55.29 | 73.14 | 80.70 | 89.53 | 93.31 |
| 175 | 42.20 | 49.47 | 73.14 | 80.70 | 89.53 | 102.14 | 103.40 |
| 200 | 42.20 | 49.47 | 73.14 | 80.70 | 89.53 | 102.14 | 103.40 |

Manhours include checking out of job storage, handling, hauling, setting, and aligning of motor as outlined.

Manhours do not include installation of supports, conduit, pulling of wire, or connecting. See respective tables for these time frames.

# MOUNTING MOTORS—VARIABLE SPEEDS TO RPM RATING LISTED, AC, 60-CYCLE, 2- AND 3-PHASE, 220, 440, OR 550 VOLTS

MANHOURS REQUIRED EACH

| Horse- | RPM Rating and Mounting Manhours | | | | | |
|---|---|---|---|---|---|---|
| power | 1750 | 1160 | 875 | 700 | 575 | 490 |
| 1 | 3.64 | 3.64 | 3.64 | 3.64 | 6.55 | 6.55 |
| 1-1/2 | 3.64 | 3.64 | 3.64 | 3.64 | 6.55 | 6.55 |
| 2 | 3.64 | 3.64 | 3.64 | 6.55 | 8.00 | 13.10 |
| 3 | 3.64 | 3.64 | 5.09 | 8.00 | 13.10 | 16.00 |
| 5 | 3.64 | 5.09 | 6.55 | 13.10 | 16.00 | 16.00 |
| 7-1/2 | 5.09 | 6.55 | 8.00 | 16.00 | 16.00 | 20.37 |
| 10 | 6.55 | 8.00 | 13.10 | 16.00 | 20.37 | 21.83 |
| 15 | 8.00 | 13.10 | 16.00 | 20.37 | 21.83 | 23.28 |
| 20 | 13.10 | 16.00 | 16.00 | 21.83 | 23.28 | 26.19 |
| 25 | 13.10 | 16.00 | 20.37 | 23.28 | 24.73 | 34.92 |
| 30 | 16.00 | 20.37 | 21.83 | 24.73 | 34.92 | 39.30 |
| 35 | – | – | 23.28 | 34.92 | 34.92 | 39.30 |
| 40 | 20.37 | 21.83 | 23.28 | 34.92 | 40.74 | 48.02 |
| 50 | 23.28 | 24.73 | 34.92 | 34.92 | 40.74 | 48.02 |
| 60 | 23.28 | 24.73 | 34.92 | 40.74 | 48.02 | 61.11 |
| 75 | 23.28 | 37.83 | 48.02 | 52.96 | 52.96 | 76.92 |
| 100 | 32.01 | 37.83 | 48.02 | 52.96 | 76.92 | 76.92 |
| 440- or 550-Volt Motors | | | | | | |
| 125 | 37.83 | 43.65 | 52.96 | 76.92 | 76.92 | 85.75 |
| 150 | 46.56 | 43.65 | 68.09 | 85.75 | 85.75 | 85.75 |
| 175 | 46.56 | 55.29 | 74.40 | 85.75 | 94.58 | 104.66 |
| 200 | 46.56 | 55.29 | 74.40 | 94.58 | 94.58 | 104.66 |

Manhours include checking out of job storage, handling, hauling, setting, and aligning of motor as outlined.

Manhours do not include installation of supports, conduit, pulling of wire, or connecting. See respective tables for these time frames.

# MOUNTING MOTORS—CONSTANT & VARIABLE SPEEDS AC, 25-CYCLE, 2- AND 3-PHASE, 220, 440, OR 550 VOLTS

MANHOURS REQUIRED EACH

| Horse-power | RPM Rating and Mounting Manhours | | | | | |
|---|---|---|---|---|---|---|
| | Constant Speeds | | | Variable Speeds | | |
| | 1440 | 720 | 475 | 1440 | 720 | 475 |
| 1 | 3.28 | 3.64 | 4.37 | 4.37 | 4.37 | 4.37 |
| 1-1/2 | 3.64 | 4.37 | 5.09 | 4.37 | 4.37 | 5.09 |
| 2 | 3.64 | 4.37 | 5.09 | 4.37 | 5.09 | 5.09 |
| 3 | 4.37 | 5.09 | 8.00 | 5.09 | 6.55 | 8.00 |
| 5 | 5.09 | 6.55 | 10.19 | 6.55 | 8.00 | 13.10 |
| 7-1/2 | 8.00 | 8.00 | 14.55 | 8.00 | 13.10 | 16.00 |
| 10 | 10.19 | 10.19 | 15.28 | 13.10 | 16.00 | 21.83 |
| 15 | 10.19 | 15.28 | 20.37 | 16.00 | 21.83 | 23.28 |
| 20 | 14.55 | 18.92 | 21.83 | 21.83 | 24.73 | 24.73 |
| 25 | 14.55 | 20.37 | 22.55 | 21.83 | 24.73 | 37.83 |
| 30 | 20.37 | 21.83 | 34.92 | 23.28 | 24.73 | 42.20 |
| 35 | 21.83 | 22.55 | 37.83 | 24.73 | 37.83 | 42.20 |
| 40 | 21.83 | 22.55 | 39.30 | 24.73 | 37.83 | 42.20 |
| 50 | 29.10 | 34.92 | 48.02 | 37.83 | 42.20 | 52.38 |
| 60 | 29.10 | 39.30 | 48.02 | 39.29 | 52.38 | 66.93 |
| 75 | 29.10 | 39.30 | 48.02 | 42.20 | 52.38 | 66.93 |
| 100 | 37.83 | 46.56 | 61.11 | 52.38 | 58.01 | 87.01 |
| 125 | 46.56 | 78.18 | 87.00 | 52.38 | 81.97 | 99.62 |
| 440- or 550-Volt Motors | | | | | | |
| 150 | 58.20 | 87.00 | 94.58 | 58.01 | 87.01 | 99.62 |
| 175 | 65.96 | 94.58 | 94.58 | 73.14 | 89.53 | 109.71 |
| 200 | 65.96 | 94.58 | 107.19 | 81.97 | 89.53 | 109.71 |

Manhours include checking out of job storage, handling, hauling, setting, and aligning of motor as outlined.

Manhours do not include installation of supports, conduit, pulling of wire, or connecting. See respective tables for these time frames.

# MOUNTING MOTORS—DC, 115-230 VOLTS, CONSTANT SPEEDS, SHUNT SERIES & COMPOUND WOUND, COMMUTATING, POLE TYPE

MANHOURS REQUIRED EACH

| Horse- | RPM Rating and Mounting Manhours | | | | | | |
|---|---|---|---|---|---|---|---|
| power | 1750 | 1150 | 850 | 690 | 575 | 500 | 450 |
| 1 | 3.28 | 3.28 | 4.37 | 5.09 | 5.09 | 6.55 | 6.55 |
| 1-1/2 | 4.37 | 4.37 | 5.09 | 5.09 | 5.09 | 6.55 | 6.55 |
| 2 | 4.37 | 4.37 | 5.09 | 6.55 | 6.55 | 10.19 | 10.19 |
| 3 | 4.37 | 5.09 | 6.55 | 10.19 | 10.19 | 10.19 | 11.64 |
| 5 | 5.09 | 6.55 | 10.19 | 10.19 | 11.64 | 14.55 | 14.55 |
| 7-1/2 | 6.55 | 10.19 | 11.64 | 11.64 | 14.55 | 16.00 | 16.00 |
| 10 | 10.19 | 11.64 | 14.55 | 14.55 | 16.00 | 21.83 | 21.83 |
| 15 | 11.64 | 15.28 | 16.00 | 21.83 | 21.83 | 23.28 | 29.10 |
| 20 | 15.28 | 16.00 | 21.83 | 21.83 | 24.73 | 29.10 | 37;83 |
| 25 | 16.00 | 21.83 | 21.83 | 24.73 | 29.10 | 37.83 | 53.84 |
| 30 | 16.00 | 21.83 | 21.83 | 29.10 | 37.83 | 53.84 | 53.84 |
| 40 | 21.83 | 21.83 | 29.10 | 37.83 | 53.84 | 53.84 | 56.75 |
| 50 | 21.83 | 24.73 | 37.83 | 53.84 | 56.75 | 56.75 | 68.09 |
| 60 | 21.83 | 29.10 | 53.84 | 56.75 | 56.75 | 68.09 | 68.09 |
| 75 | 24.73 | 37.83 | 56.75 | 66.93 | 68.09 | 68.09 | 85.75 |
| 100 | 37.83 | 53.84 | 56.75 | 78.57 | 85.75 | 94.58 | 100.88 |
| 125 | 53.84 | 56.75 | 68.09 | 85.75 | 94.58 | 108.45 | 115.11 |
| 150 | 56.75 | 68.09 | 87.00 | 106.80 | 108.45 | 115.11 | 117.28 |
| 175 | 68.09 | 87.00 | 97.10 | 108.45 | 118.53 | 119.80 | 122.32 |
| 200 | 87.00 | 97.10 | 108.45 | 118.53 | 119.80 | 122.32 | 126.10 |

Manhours include checking out of job storage, handling, hauling, setting, and aligning of motor as outlined.

Manhours do not include installation of supports, conduit, pulling of wire, or connecting. See respective tables for these time frames.

# Section 7

# PAINTING

Painting is usually considered a specialty item throughout the construction industry and as such is usually subcontracted to a contractor who specializes in this trade. This section provides the mechanical estimator a sound means of estimating painting labor as may be involved with the plumbing, heating, ventilating, or air-conditioning portion of an industrial or chemical plant. These tables may be used as the primary means of labor estimation or as a check estimate to analyze the subcontractor's price.

This section does not cover pipe color coding because the scope of the work involved in this operation can vary so greatly. As an example, at one setup location as many as a dozen lines may be banded, but on the other hand, the same setup may be required to band only one line. Therefore, each operation must be individually examined according to piping specifications and locations.

There has been no attempt to outline the method or procedure that an estimator should use in taking off this type of work. These manhour tables are based on labor estimates made in accordance with the standard measurements used throughout the painting and decorating trades. The estimator should understand the various methods of measurement and measurement allowances that are used in the taking off of specialty items.

# EXTERIOR IRON & STEEL

MANHOURS PER UNITS LISTED

| Item | Unit | Painter Manhours | | | |
|---|---|---|---|---|---|
| | | Prime Coat | First Coat | Add. Coat | Total Coats |
| Structural Steel | | | | | |
| Sandblast, scale & derust | ton | – | – | 1.00 | 1.00 |
| Wire brush, scale & derust | ton | – | – | 5.20 | 5.20 |
| Shop coat | ton | .70 | – | .70 | .70 |
| Brush field coat (erected) | 100 sq ft | – | 1.00 | .67 | 1.67 |
| Spray coat (erected) | 100 sq ft | – | .22 | .15 | .37 |
| Miscellaneous Iron | ton | – | 1.60 | 1.40 | 3.00 |
| Metal Surfaces | 100 sq ft | – | .48 | .40 | .88 |
| Metal Trim | 100 sq ft | – | .88 | .85 | 1.73 |
| Metal Deck | 100 sq ft | – | .52 | .50 | 1.02 |
| Steel Sash (both sides) | 100 sq ft | – | 1.78 | 1.75 | 3.53 |
| Metal Doors | 100 sq ft | – | .56 | .53 | 1.09 |

Manhours include handling, stirring, mixing and placing of paint on items as outlined above.

Additional coat manhours are for one coat only. Add same number of manhours for each additional coat applied.

Manhours for sandblasting are those of air tool operator; for wire brushing – those of laborer.

Manhours do not include scaffolding. See respective table for this charge.

# BRUSH PAINT EQUIPMENT

MANHOURS PER SQUARE FOOT

| Surface Area or Item | Manhours |
|---|---|
| Small flat areas | .020 |
| Large flat areas | .014 |
| Machinery | .033 |

Manhours include handling, stirring, mixing, and brushing on one coat.

For second coat decrease manhours 20% per square foot.

# BRUSH INTERIOR – METAL WORK

MANHOURS PER UNITS LISTED

| Item | Unit | Painter Manhours | | |
|---|---|---|---|---|
| | | Prime Coat | Second Coat | Total Coats |
| Miscellaneous Iron | ton | 1.68 | 2.12 | 3.80 |
| Metal Surfaces | 100 sq ft | .56 | .64 | 1.20 |
| Metal Trim | 100 sq ft | .80 | .88 | 1.68 |
| Metal Doors | 100 sq ft | .60 | .88 | 1.48 |

Manhours include handling, stirring and mixing, and brushing on of red lead or similar type of prime coat and finish second coat.

If third coat is required, add manhours for second coat to above total.

Manhours do not include scaffolding. See respective table for this charge.

# SAND BLAST & PAINT PIPE

## Commercial Blast

MANHOURS PER LINEAL FOOT

| Nominal Size | 4 Coats Conventional Paint | 4 Coats Chlorinated Rubber | 4 Coats Vinyl Paint | 1 Coat Dimetcote #3 | 5 Coats Epoxy Paint | 1/16" Barretts 10–70 |
|---|---|---|---|---|---|---|
| 2 | 0.05 | 0.05 | 0.06 | 0.05 | 0.08 | 0.04 |
| 2-1/2 | 0.05 | 0.06 | 0.08 | 0.07 | 0.10 | 0.05 |
| 3 | 0.06 | 0.07 | 0.09 | 0.08 | 0.12 | 0.06 |
| 3-1/2 | 0.07 | 0.08 | 0.10 | 0.09 | 0.13 | 0.07 |
| 4 | 0.08 | 0.08 | 0.10 | 0.10 | 0.14 | 0.07 |
| 5 | 0.09 | 0.10 | 0.13 | 0.11 | 0.17 | 0.08 |
| 6 | 0.10 | 0.12 | 0.15 | 0.13 | 0.19 | 0.10 |
| 8 | 0.13 | 0.15 | 0.19 | 0.17 | 0.24 | 0.13 |
| 10 | 0.16 | 0.18 | 0.23 | 0.20 | 0.29 | 0.15 |
| 12 | 0.18 | 0.19 | 0.25 | 0.21 | 0.32 | 0.16 |
| 14 | 0.19 | 0.21 | 0.27 | 0.24 | 0.35 | 0.18 |
| 16 | 0.22 | 0.24 | 0.31 | 0.27 | 0.40 | 0.21 |
| 18 | 0.25 | 0.27 | 0.35 | 0.31 | 0.45 | 0.24 |
| 20 | 0.28 | 0.31 | 0.40 | 0.34 | 0.50 | 0.26 |
| 24 | 0.34 | 0.37 | 0.47 | 0.42 | 0.60 | 0.32 |
| 30 | 0.42 | 0.46 | 0.59 | 0.52 | 0.75 | 0.40 |
| 36 | 0.50 | 0.56 | 0.71 | 0.63 | 0.90 | 0.48 |
| 42 | 0.59 | 0.65 | 0.83 | 0.74 | 1.06 | 0.56 |
| 48 | 0.65 | 0.72 | 0.91 | 0.81 | 1.16 | 0.61 |
| 60 | 0.85 | 0.93 | 1.19 | 1.05 | 1.51 | 0.80 |

Manhours for painting pipe only. Labor for scaffolding must be added.

Manhours for galvanizing exterior of pipe only is approximately 80% of conventional paint.

Manhours to galvanize exterior and interior of pipe is approximately the same as dimetcote.

# Section 8

## EARTHWORK

This section covers as nearly as possible all operations pertaining to the removal and replacement of earth materials for the installation of underground piping, septic tanks, and drainage structures.

The following manhours cover labor only for the various operations as outlined in the individual tables and in accordance with the notes thereon.

No emphasis has been placed on the degree of swing or depth of cut as may pertain to the efficiency rating of a piece of heavy equipment such as a power shovel, dragline, etc. The following manhours are average for this type of work. Should a particular type of earthwork of a very difficult nature be encountered, the estimator should give this due consideration and adjust accordingly.

Before an estimate is made on excavation, it is well to know the kind of soil that may be encountered. For this reason we have divided soil into five groups according to the difficulty experienced in excavating it. Soils vary greatly in character and no two are exactly alike.

Group 1: LIGHT SOIL – Earth which can be shoveled easily and requires no loosening, such as sand.

Group 2: MEDIUM OR ORDINARY SOILS – Type of earth easily loosened by pick. Preliminary loosening is not required when power excavating equipment such as shovels, dragline scrapers, and backhoes are used. This earth is usually classified as ordinary soil and loam.

Group 3: HEAVY OR HARD SOIL – This type of soil can be loosened by pick but this loosening is sometimes very hard to do. It may be exacavated by sturdy power shovels without preliminary loosening. Hard and compact loam containing gravel, small stones and boulders, stiff clay, or compacted gravel are good examples of this type.

Group 4: HARD PAN OR SHALE – A soil that has hardened and is very difficult to loosen with picks. Light blasting is often required when exacavating with power equipment.

Group 5: ROCK – Requires blasting before removal and transporting. May be divided into different grades such as hard, soft, or medium.

From the following manhour tables, a complete direct labor estimate can be made for excavating almost any type of soil which may be encountered.

# HAND EXCAVATION

MANHOURS PER CUBIC YARD

| Soil | Excavation | MANHOURS | | |
|---|---|---|---|---|
| | | First Lift | Second Lift | Third Lift |
| Light | General Dry | 1.07 | 1.42 | 1.89 |
| | General Wet | 1.60 | 2.13 | 2.83 |
| | Special Dry | 1.34 | 1.78 | 3.37 |
| Medium | General Dry | 1.60 | 2.19 | 2.83 |
| | General Wet | 2.14 | 2.85 | 3.79 |
| | Special Dry | 2.00 | 2.49 | 3.31 |
| Hard or Heavy | General Dry | 2.67 | 3.55 | 4.72 |
| | General Wet | 3.21 | 4.27 | 5.68 |
| | Special Dry | 2.94 | 3.91 | 5.70 |
| Hard Pan | General Dry | 3.74 | 4.97 | 6.61 |
| | General Wet | 4.28 | 5.69 | 7.57 |
| | Special Dry | 4.01 | 5.33 | 7.09 |

Manhours include picking and loosening where necessary and placing on bank out of way of excavation or loading into trucks or wagons for hauling away.

Manhours do not include blasting, hauling or unloading. See respective tables for these charges.

# HAND SHAPE TRENCH BOTTOM

MANHOURS PER HUNDRED (100) LINEAR FEET

| Item | Manhours |
|---|---|
| | Laborers |
| Shape Trench for Drain Pipes | |
| 6" to 10" Pipe | 2.25 |
| 12" to 15" Pipe | 5.25 |
| 18" to 21" Pipe | 7.50 |
| 24" Pipe | 13.00 |
| 30" Pipe | 22.10 |
| 36" Pipe and Larger | 28.60 |

Above manhours include minor hand excavations for grade and excavations for bells where required for all types of drain lines such as tile, concrete and corrugated metal in pre-excavated ditches.

Manhours are average for one hundred (100) linear feet of ditch measured on center line.

Manhours do not include excavation of ditch or placement of pipes. See respective tables for these charges.

# MACHINE EXCAVATION – POWER SHOVEL

MANHOURS PER HUNDRED (100) CUBIC YARDS

| Soil | Dipper Size | Manhours | | | |
|---|---|---|---|---|---|
| | | Oper. Engr. | Oiler | Laborer | Total |
| Light | 1 cubic yard dipper | 1.1 | 1.1 | 1.1 | 3.3 |
| | ¾ cubic yard dipper | 1.5 | 1.5 | 1.5 | 4.5 |
| | ½ cubic yard dipper | 2.0 | 2.0 | 2.0 | 6.0 |
| Medium | 1 cubic yard dipper | 2.0 | 2.0 | 2.0 | 6.0 |
| | ¾ cubic yard dipper | 2.8 | 2.8 | 2.8 | 8.4 |
| | ½ cubic yard dipper | 3.7 | 3.7 | 3.7 | 11.1 |
| Heavy | 1 cubic yard dipper | 2.7 | 2.7 | 2.7 | 8.1 |
| | ¾ cubic yard dipper | 3.7 | 3.7 | 3.7 | 11.1 |
| | ½ cubic yard dipper | 4.9 | 4.9 | 4.9 | 14.7 |
| Hard Pan | 1 cubic yard dipper | 3.4 | 3.4 | 3.4 | 10.2 |
| | ¾ cubic yard dipper | 4.6 | 4.6 | 4.6 | 13.8 |
| | ½ cubic yard dipper | 6.1 | 6.1 | 6.1 | 18.3 |
| Rock | 1 cubic yard dipper | 3.4 | 3.4 | 3.4 | 10.2 |
| | ¾ cubic yard dipper | 4.6 | 4.6 | 4.6 | 13.8 |
| | ½ cubic yard dipper | 6.1 | 6.1 | 6.1 | 18.3 |

Manhours include operations of excavating and dumping on side lines or loading into trucks.

Manhours do not include hauling or blasting. See respective tables for these charges.

Above manhours are based on excavations up to six (6) feet. If excavations are to be greater in depth than this, the estimator should consider additional methods, planning, and equipment required.

If total excavated quantity is less than one hundred (100) cubic yards increase above units by thirty (30) percent.

# MACHINE EXCAVATION – BACK HOE

MANHOURS PER HUNDRED (100) CUBIC YARDS

| Soil | Bucket Size | Manhours | | | |
|---|---|---|---|---|---|
| | | Oper. Engr. | Oiler | Laborer | Total |
| Light | 1 cubic yard bucket | 1.4 | 1.4 | 1.4 | 4.2 |
| | ¾ cubic yard bucket | 1.5 | 1.5 | 1.5 | 4.5 |
| | ½ cubic yard bucket | 2.0 | 2.0 | 2.0 | 6.0 |
| Medium | 1 cubic yard bucket | 2.6 | 2.6 | 2.6 | 7.8 |
| | ¾ cubic yard bucket | 3.8 | 3.8 | 3.8 | 11.4 |
| | ½ cubic yard bucket | 4.4 | 4.4 | 4.4 | 13.2 |
| Heavy | 1 cubic yard bucket | 3.5 | 3.5 | 3.5 | 10.5 |
| | ¾ cubic yard bucket | 4.0 | 4.0 | 4.0 | 12.0 |
| | ½ cubic yard bucket | 4.9 | 4.9 | 4.9 | 14.7 |
| Hard Pan | 1 cubic yard bucket | 4.4 | 4.4 | 4.4 | 13.2 |
| | ¾ cubic yard bucket | 4.6 | 4.6 | 4.6 | 13.8 |
| | ½ cubic yard bucket | 6.1 | 6.1 | 6.1 | 18.3 |
| Rock | 1 cubic yard bucket | 4.4 | 4.4 | 4.4 | 13.2 |
| | ¾ cubic yard bucket | 4.6 | 4.6 | 4.6 | 13.8 |
| | ½ cubic yard bucket | 6.1 | 6.1 | 6.1 | 18.3 |

Manhours include operations of excavating and dumping on side lines or loading into trucks.

Manhours do not include hauling or blasting. See respective tables for these charges.

Above manhours are based on excavations up to six (6) feet in depth. If excavations are to be greater in depth than this the estimator should consider methods, planning and equipment required.

If total excavation quantity is less than one hundred (100) cubic yards, increase above units by thirty (30) percent.

# MACHINE EXCAVATION – TRENCHING MACHINE & DRAGLINE

MANHOURS PER HUNDRED (100) CUBIC YARDS

| Soil | Item | Manhours | | | |
|---|---|---|---|---|---|
| | | Oper. Engr. | Oiler | Laborer | Total |
| | DRAGLINE: | | | | |
| Light | 2 cubic yard bucket | 0.7 | 0.7 | 0.7 | 2.1 |
| | 1 cubic yard bucket | 1.1 | 1.1 | 1.1 | 3.3 |
| | ½ cubic yard bucket | 2.0 | 2.0 | 2.0 | 6.0 |
| Medium | 2 cubic yard bucket | 1.3 | 1.3 | 1.3 | 3.9 |
| | 1 cubic yard bucket | 2.0 | 2.0 | 2.0 | 6.0 |
| | ½ cubic yard bucket | 3.7 | 3.7 | 3.7 | 11.1 |
| Heavy | 2 cubic yard bucket | 1.7 | 1.7 | 1.7 | 5.1 |
| | 1 cubic yard bucket | 2.7 | 2.7 | 2.7 | 8.1 |
| | ½ cubic yard bucket | 4.9 | 4.9 | 4.9 | 14.7 |
| Medium | TRENCHING MACHINE: | 3.8 | – | 7.5 | 11.3 |
| Heavy | TRENCHING MACHINE: | 4.8 | – | 9.4 | 14.2 |

Manhours include operations of excavating and dumping on side lines or loading into trucks for dragline excavation.

Manhours for trenching machine include regular trenching up to 3'-6" wide.

Manhours do not include hauling. See respective tables for this charge.

Above manhours are based on excavating up to six (6) feet deep. If excavations are to be greater in depth than this, the estimator should consider additional methods, planning, and equipment required.

If total excavated quantity is less than one hundred (100) cubic yards, increase above manhours by thirty (30) percent.

# ROCK EXCAVATION

MANHOURS PER UNITS LISTED

| Type Rock | Operation | Unit | Manhours | | | | |
|---|---|---|---|---|---|---|---|
| | | | Laborer | Air Tool Operator | Oper. Engr. | Powder Man | Total |
| Soft | Drilling Holes | | | | | | |
| | 2½" with jackhammer | lin ft | .06 | .06 | .06 | — | .18 |
| | 2" with jackhammer | lin ft | .04 | .04 | .04 | — | .12 |
| | 2" with wagon drill | lin ft | .01 | — | .07 | — | .08 |
| | Blasting | cu yd | .04 | — | — | .02 | .06 |
| Medium | Drilling Holes | | | | | | |
| | 2½" with jackhammer | lin ft | .08 | .08 | .08 | — | .24 |
| | 2" with jackhammer | lin ft | .07 | .07 | .07 | — | .21 |
| | 2" with wagon drill | lin ft | .03 | — | .10 | — | .13 |
| | Blasting | cu yd | .06 | — | — | .02 | .08 |
| Hard | Drilling Holes | | | | | | |
| | 2½" with jackhammer | lin ft | .10 | .10 | .10 | — | .30 |
| | 2" with jackhammer | lin ft | .09 | .09 | .09 | — | .27 |
| | 2" with wagon drill | lin ft | .05 | — | .15 | — | .20 |
| | Blasting | cu yd | .09 | — | — | .04 | .13 |

Manhours are for the above operations for primary blasting only. If secondary blasting is required and this same method is used, increase above manhours 50 percent. If heavy weight or headache ball and crane are used for secondary breakage, refer to table under demolition for breakage of concrete slabs using this method.

Manhours do not include loading or hauling of blasted materials. See respective tables for these charges.

# DISPOSAL OF EXCAVATED MATERIALS

MANHOURS PER HUNDRED (100) CUBIC YARDS

| Truck Capacity and Length of Haul | Manhours | | | | | | | | |
|---|---|---|---|---|---|---|---|---|---|
| | Average Speed 10 mph | | | Average Speed 15 mph | | | Average Speed 20 mph | | |
| | Truck Driver | Laborer | Total | Truck Driver | Laborer | Total | Truck Driver | Laborer | Total |
| 3 Cu Yd Truck: | | | | | | | | | |
| 1 Mile Haul | 15.0 | 2.8 | 17.8 | 11.6 | 2.8 | 14.4 | 10.5 | 2.8 | 13.3 |
| 2 Mile Haul | 21.8 | 2.8 | 24.6 | 16.2 | 2.8 | 19.0 | 14.0 | 2.8 | 16.8 |
| 3 Mile Haul | 28.2 | 3.0 | 31.2 | 20.6 | 3.0 | 23.6 | 17.3 | 3.0 | 20.3 |
| 4 Mile Haul | 36.0 | 3.0 | 39.0 | 26.8 | 3.0 | 29.8 | 21.0 | 3.0 | 24.0 |
| 5 Mile Haul | 41.7 | 2.5 | 44.2 | — | — | — | — | — | — |
| 4 Cu Yd Truck: | | | | | | | | | |
| 1 Mile Haul | 11.3 | 2.1 | 13.4 | 8.8 | 2.0 | 10.8 | 7.9 | 2.1 | 9.0 |
| 2 Mile Haul | 16.2 | 2.1 | 18.3 | 12.0 | 2.0 | 14.0 | 10.4 | 2.1 | 12.5 |
| 3 Mile Haul | 21.6 | 2.0 | 23.6 | 15.8 | 2.3 | 18.1 | 13.2 | 2.2 | 15.4 |
| 4 Mile Haul | 26.4 | 2.0 | 28.4 | 18.7 | 2.3 | 21.0 | 15.6 | 2.2 | 17.8 |
| 5 Mile Haul | 31.3 | 1.3 | 32.6 | — | — | — | — | — | — |
| 5 Cu Yd Truck: | | | | | | | | | |
| 1 Mile Haul | 9.0 | 1.7 | 10.7 | 7.0 | 1.7 | 8.7 | 6.3 | 1.6 | 7.9 |
| 2 Mile Haul | 13.0 | 1.7 | 14.7 | 9.7 | 1.7 | 11.4 | 8.3 | 1.7 | 10.0 |
| 3 Mile Haul | 17.1 | 1.8 | 18.9 | 12.3 | 1.8 | 14.1 | 10.4 | 1.7 | 12.1 |
| 4 Mile Haul | 21.0 | 2.0 | 23.0 | 15.0 | 2.0 | 17.0 | 12.4 | 1.7 | 14.1 |
| 5 Mile Haul | 25.0 | 1.7 | 26.7 | — | — | — | — | — | — |
| 8 Cu Yd Truck: | | | | | | | | | |
| 1 Mile Haul | 5.6 | 1.0 | 6.6 | 4.8 | 1.0 | 5.8 | 4.0 | 1.0 | 5.0 |
| 2 Mile Haul | 8.2 | 1.0 | 9.2 | 6.0 | 1.0 | 7.0 | 5.2 | 1.0 | 6.2 |
| 3 Mile Haul | 10.5 | 1.1 | 11.6 | 7.8 | 1.1 | 8.9 | 6.5 | 1.0 | 7.5 |
| 4 Mile Haul | 13.2 | 1.1 | 14.3 | 9.2 | 1.1 | 10.3 | 7.6 | 1.0 | 8.6 |
| 5 Mile Haul | 15.6 | 1.3 | 16.9 | — | — | — | — | — | — |

Manhours include round trip for truck driver, spotting at both ends, unloading and labor for minor repairs.

Manhours do not include labor for excavation or loading of trucks. See respective tables for these charges.

# DISPOSAL OF EXCAVATED MATERIALS

MANHOURS PER HUNDRED (100) CUBIC YARDS

| Truck Capacity and Length of Haul | MANHOURS | | | | | | | | |
|---|---|---|---|---|---|---|---|---|---|
| | Average Speed 20 mph | | | Average Speed 25 mph | | | Average Speed 30 mph | | |
| | Truck Driver | Laborer | Total | Truck Driver | Laborer | Total | Truck Driver | Laborer | Total |
| 3 Cu Yd Truck: | | | | | | | | | |
| 6 Mile Haul | 26.0 | 2.5 | 28.5 | 21.4 | 2.5 | 23.9 | 17.6 | 2.4 | 20.0 |
| 7 Mile Haul | 27.1 | 2.3 | 29.4 | 22.3 | 2.3 | 24.6 | 18.3 | 2.2 | 20.5 |
| 8 Mile Haul | 28.7 | 2.3 | 31.0 | 23.6 | 2.3 | 25.9 | 19.4 | 2.2 | 21.6 |
| 9 Mile Haul | 30.8 | 2.1 | 32.9 | 25.3 | 2.1 | 27.4 | 20.8 | 2.0 | 22.8 |
| 10 Mile Haul | 33.4 | 2.1 | 35.5 | 27.5 | 2.1 | 29.6 | 22.6 | 2.0 | 24.6 |
| 4 Cu Yd Truck: | | | | | | | | | |
| 6 Mile Haul | 19.3 | 1.5 | 20.8 | 16.0 | 2.0 | 18.0 | 13.4 | 1.9 | 15.3 |
| 7 Mile Haul | 21.0 | 1.5 | 22.5 | 17.4 | 2.0 | 19.4 | 14.8 | 1.9 | 16.7 |
| 8 Mile Haul | 23.5 | 1.5 | 25.0 | 19.5 | 1.8 | 21.3 | 16.6 | 1.8 | 18.4 |
| 9 Mile Haul | 26.9 | 1.3 | 28.2 | 22.3 | 1.8 | 24.1 | 19.0 | 1.8 | 20.8 |
| 10 Mile Haul | 31.1 | 1.3 | 32.4 | 25.8 | 1.6 | 27.4 | 21.9 | 1.5 | 23.4 |
| 5 Cu Yd Truck: | | | | | | | | | |
| 6 Mile Haul | 15.6 | 1.4 | 17.0 | 14.1 | 1.6 | 15.7 | 11.7 | 1.5 | 13.2 |
| 7 Mile Haul | 17.3 | 1.4 | 18.7 | 15.5 | 1.5 | 17.0 | 12.9 | 1.4 | 14.3 |
| 8 Mile Haul | 19.8 | 1.3 | 21.1 | 17.5 | 1.5 | 19.0 | 14.5 | 1.3 | 15.8 |
| 9 Mile Haul | 23.2 | 1.3 | 24.5 | 20.3 | 1.2 | 21.5 | 16.8 | 1.1 | 17.9 |
| 10 Mile Haul | 27.4 | 1.2 | 28.6 | 23.7 | 1.2 | 24.9 | 19.7 | 1.1 | 20.8 |
| 8 Cu Yd Truck: | | | | | | | | | |
| 6 Mile Haul | 9.8 | 1.2 | 11.0 | 9.2 | 1.0 | 10.2 | 7.7 | 1.0 | 8.7 |
| 7 Mile Haul | 11.5 | 1.2 | 12.7 | 10.6 | 1.0 | 11.6 | 8.9 | 1.0 | 9.9 |
| 8 Mile Haul | 14.1 | 1.1 | 15.2 | 12.7 | 1.0 | 13.7 | 10.6 | 0.9 | 11.5 |
| 9 Mile Haul | 17.4 | 1.1 | 18.5 | 15.4 | 0.9 | 16.3 | 12.9 | 0.9 | 13.8 |
| 10 Mile Haul | 21.7 | 1.0 | 22.7 | 19.0 | 0.9 | 19.9 | 15.9 | 0.8 | 16.7 |

Manhours include round trip for truck driver, spotting at both ends, unloading, and labor for minor repairs.

Manhours do not include labor for excavation or loading of trucks. See respective tables for these charges.

# LOADING DIRT FROM STOCKPILE WITH CLAMSHELL

MANHOURS PER HUNDRED (100) CUBIC YARDS

| Item | Manhours | | |
|---|---|---|---|
| | Oper. Engr. | Oiler | Total |
| Clamshell | | | |
| 2 cubic yard bucket | 1.10 | 1.10 | 2.20 |
| 1 cubic yard bucket | 2.00 | 2.00 | 4.00 |
| ¾ cubic yard bucket | 2.50 | 2.50 | 5.00 |
| ½ cubic yard bucket | 3.30 | 3.30 | 6.60 |

Manhours are for loading dirt from stock pile into trucks for hauling, using the type of equipment as outlined above.

Manhours do not include spotting trucks or hauling. See respective tables for these charges.

# MACHINE & HAND BACKFILL

**Average for Sand or Loam, Ordinary Soil, Heavy Soil and Clay**

MANHOURS PER UNITS LISTED

| Item | Unit | Manhours | | | |
|---|---|---|---|---|---|
| | | Laborer | Oper. Engr. | Oiler | Total |
| Hand Place | cu yd | .55 | – | – | .55 |
| Bulldoze Loose Material | 100 cu yds | – | 3.32 | – | 3.32 |
| Clamshell | | | | | |
| 1 cubic yard bucket | 100 cu yds | – | 1.60 | 1.60 | 3.20 |
| ¾ cubic yard bucket | 100 cu yds | – | 2.00 | 2.00 | 4.00 |
| ½ cubic yard bucket | 100 cu yds | – | 2.75 | 2.75 | 5.50 |
| Hand Spread | | | | | |
| Stone or gravel fill | cu yd | .40 | – | – | .40 |
| Sand fill | cu yd | .35 | – | – | .35 |
| Cinder fill | cu yd | .40 | – | – | .40 |
| Tamp by Hand | cu yd | .60 | – | – | .60 |
| Pneumatic Tamping | cu yd | .25 | – | – | .25 |

Hand Place units include: Placing by hand with shovels loose earth within hand-trowing distance of stockpiles. This unit does not include compaction.

Bulldoze Loose Material units include: The moving of pre-stockpiled loose earth over an area.

Clamshell units include: The placement of materials from reachable stockpiles.

Stone, Sand and Cinder Spread units include: The hand placing, with shovels, these materials from strategically located stockpiles.

Tamp By Hand and Pneumatic Tamping units include: The compacting of pre-spread materials in 6" layers. Manhours above for this type work are shown as laborer hours. Should air tool operator be required for this work – substitute his time for above laborer hours.

Manhours do not include trucking or fine grading. See respective tables for these charges.

# HAND PLACED WOOD SHEET PILING, SHORING & BRACING

MANHOURS PER HUNDRED (100) SQUARE FEET

| Item | Manhours | | | |
|---|---|---|---|---|
| | Carpenter | Laborer | Truck Driver | Total |
| Sheet Piling - Basements & Pits | | | | |
| Placing | 6.75 | 6.75 | .30 | 13.80 |
| Removing | — | 4.33 | .30 | 4.63 |
| Sheet Piling - Trenches to 8' Deep | | | | |
| Placing | 5.00 | 5.00 | .30 | 10.30 |
| Removing | — | 2.50 | .30 | 2.80 |
| Sheet Piling - Trenches over 8' Deep | | | | |
| Placing | 5.55 | 5.55 | .30 | 11.40 |
| Removing | — | 3.20 | .30 | 3.50 |
| Shoring or Bracing Trenches | | | | |
| Placing | 2.50 | 2.50 | .20 | 5.20 |
| Removing | — | 1.50 | .20 | 1.70 |

Manhours include necessary labor as may be involved for the hauling, handling, fabricating, placing and removing wood sheet piling, shoring and bracing as outlined above.

Manhours do not include excavation or pumping. See respective tables for these charges.

# PNEUMATIC DRIVEN WOOD SHEET PILING, SHORING & BRACING

MANHOURS PER HUNDRED (100) SQUARE FEET

| Item | Manhours | | | | | |
|---|---|---|---|---|---|---|
| | Carpenter | Laborer | Oper. Engr. | Air Tool Operator | Truck Driver | Total |
| **Sheet Piling for Pits** | | | | | | |
| Placing | 2.80 | 2.80 | 2.80 | 2.80 | .30 | 11.50 |
| Removing | — | 4.00 | — | — | .30 | 4.30 |
| **Sheet Piling - Trenches to 8' Deep** | | | | | | |
| Placing | 1.80 | 1.80 | 1.80 | 1.80 | .30 | 7.50 |
| Removing | — | 2.75 | — | — | .30 | 3.05 |
| **Sheet Piling - Trenches over 8' Deep** | | | | | | |
| Placing | 2.10 | 2.10 | 2.10 | 2.10 | .30 | 8.70 |
| Removing | — | 3.00 | — | — | .30 | 3.30 |

Manhours include necessary labor for the handling, hauling, fabricating, placing and removing wood sheet piling shored and braced in place, using a pneumatic hammer fed by compressed air.

Manhours do not include excavating or pumping. See respective tables for these charges.

# Section 9

# CONCRETE & MASONRY WORK

It is not the intent of this section to cover all types of concrete and masonry work, but only those types related to plumbing and drainage work.

The following manhour tables are average of many projects under varied conditions where strict methods and planning were followed along with strict reporting and recording of actual costs.

The listed manhours include time allowance to complete all necessary labor for the particular operation as may be outlined in the various tables and in accordance with the notes thereon.

# WOOD MANHOLE & SEPTIC TANK FORMS

MANHOUR PER SQUARE FOOT

| Item | MANHOURS | | | |
|---|---|---|---|---|
| | Carpenter | Laborer | Truck Driver | Total |
| Square Type | | | | |
| Slab on Ground | | | | |
| Fabricate | .04 | .01 | – | .05 |
| Erect | .03 | .02 | .01 | .06 |
| Strip & clean | .01 | .02 | .01 | .04 |
| Total | .08 | .05 | .02 | .15 |
| Walls | | | | |
| Fabricate | .04 | .01 | – | .05 |
| Erect | .05 | .01 | .01 | .07 |
| Strip & Clean | .01 | .02 | .01 | .04 |
| Total | .10 | .04 | .02 | .16 |
| Elevated slab with Wood shores | | | | |
| Make-up & erect | .07 | .04 | .01 | .12 |
| Strip & Clean | .01 | .04 | .01 | .06 |
| Total | .08 | .08 | .02 | .18 |

Manhours are based on the fabrication and installation of one and two-inch materials for the type of formwork as outlined above.

Manhours do not include placement of reinforcement steel or pouring concrete. See respective tables for these charges.

# BOX TYPE CULVERT & THRUST BLOCK FORMS

MANHOURS PER SQUARE FOOT

| Item | MANHOURS | | | |
|---|---|---|---|---|
| | Carpenter | Laborer | Truck Driver | Total |
| Box Type Culverts | | | | |
| Slab on Ground | .09 | .04 | .01 | .14 |
| Walls | .08 | .03 | .01 | .12 |
| Wing Walls | .10 | .02 | .01 | .13 |
| Elevated Slabs | .12 | .05 | .01 | .18 |
| Strip & clean | .01 | .02 | .01 | .04 |
| Thrust Blocks | .10 | .03 | .01 | .14 |

Manhours are based on the use of one- and two-inch wood material for the fabrication and erection of forms for the work described above

Manhours do not include placement of reinforcing steel or pouring concrete. See respective tables for these charges.

# WOOD FORMS FOR EQUIPMENT FOUNDATIONS–SIMPLE LAYOUT

MANHOURS PER SQUARE FOOT

| Item | Manhours | | | |
|---|---|---|---|---|
| | Carpenter | Laborer | Truck Driver | Total |
| Square Pads 6" to 18" High Ground Floor | | | | |
| Build in place | .14 | .03 | .01 | .18 |
| Strip & clean | .01 | .03 | – | .04 |
| Total | .15 | .06 | .01 | .22 |
| Square Pads 6" to 18" High Elevated Floors | | | | |
| Build in place | .16 | .03 | .01 | .20 |
| Strip & clean | .01 | .04 | – | .05 |
| Total | .17 | .07 | .01 | .25 |
| Square Pads to 4' High Ground Floor | | | | |
| Fabricate & erect | .17 | .04 | .01 | .22 |
| Strip & clean | .02 | .04 | – | .06 |
| Total | .19 | .08 | .01 | .28 |
| Square Pads to 4' High Elevated Floors | | | | |
| Fabricate & erect | .19 | .04 | .01 | .24 |
| Strip & clean | .02 | ,05 | – | .07 |
| Total | .21 | .09 | .01 | .31 |

Manhours are based on the fabrication and installation of two-inch materials for formwork to eighteen inches high, and plywood sheathing for forms to four feet high, all properly braced and anchored in place.

A simple layout is that of a small square pad poured either integrally with floor or over pre-set dowels left purposely in pre-poured floor for this reason.

Manhours do not include the placement or setting of anchor bolts or miscellaneous embedded steel items. See respective tables for these charges.

# WOOD FORMS FOR EQUIPMENT FOUNDATIONS COMPLEX LAYOUT

**Bulky, Offset, Skewed and Angled**

MANHOURS PER SQUARE FOOT

| Item | Manhours | | | |
|---|---|---|---|---|
| | Carpenter | Laborer | Truck Driver | Total |
| Average All Heights & Sizes | | | | |
| Fabricate & erect | .20 | .08 | .02 | .30 |
| Strip & clean | .05 | .12 | – | .17 |
| Total | .25 | .20 | .02 | .47 |
| Tank Cradle Forms | | | | |
| Build in place | .13 | .04 | .01 | .19 |
| Strip & clean | .01 | .03 | .01 | .04 |
| Total | .14 | .07 | .02 | .23 |

**Complex Foundation manhours are average for all sizes and shapes and are based on the use of one- and two-inch planking, plywood sheathing and minor sheet metal cuts and bends.**

**A complex layout is that of a large and bulky foundation with many offsets, skews and angles, such as a foundation for a turbo-generator, etc.**

**Manhours do not include the placement or setting of anchor bolts or miscellaneous embedded steel items. See respective tables for these charges.**

# HOOK TYPE ANCHOR BOLT INSTALLATION

MANHOURS EACH

| Size | Manhours Each For Overall Length | | | | | | | |
|---|---|---|---|---|---|---|---|---|
| | 0' 8" | 1' 0" | 1' 6" | 2' 0" | 2' 6" | 3' 0" | 3' 6" | 4' 0" |
| 1/4" | .15 | .15 | .20 | .20 | – | – | – | – |
| 3/8" | .15 | .15 | .20 | .22 | – | – | – | – |
| 1/2" | .15 | .15 | .25 | .28 | – | – | – | – |
| 5/8" | .15 | .20 | .25 | .28 | .30 | .33 | – | – |
| 3/4" | .18 | .20 | .28 | .30 | .35 | .38 | – | – |
| 7/8" | – | – | .40 | .43 | .45 | .48 | .50 | .53 |
| 1" | – | – | .40 | .45 | .48 | .50 | .53 | .58 |
| 1-1/4" | – | – | .48 | .50 | .50 | .53 | .55 | .65 |
| 1-1/2" | – | – | .50 | .55 | .55 | .58 | .60 | .70 |
| 1-3/4" | – | – | .55 | .58 | .60 | .65 | .68 | .73 |
| 2" | – | – | – | .65 | .68 | .70 | .75 | .78 |
| 2-1/4" | – | – | – | .70 | .73 | .75 | .78 | .80 |
| 2-1/2" | – | – | – | .75 | .78 | .78 | .80 | .85 |

Manhours are based on overall length of anchor bolt from end to end including hook and are average for all heights.

Manhours are for installation of template and bolt, or bolt and sleever, as the case may be, for the size and length as outlined above.

All bolts 7/8" and larger are assumed to be sleeved and those smaller than 7/8" round are assumed to be without sleeves.

When converting manhours to labor dollars, consideration should be given to the type of crew or crafts involved in the installation of above anchor bolts and a crew composite rate applied accordingly.

Manhours do not include fabrication of bolts.

Manhours do not include field engineering time spent aligning and checking bolts. This is usually a part of field overhead and should be considered as such.

For sizes not listed, take the next highest listing.

# HOOK TYPE ANCHOR BOLT INSTALLATION

MANHOURS EACH

| Size | Manhours each For Overall Length | | | | | | | |
|---|---|---|---|---|---|---|---|---|
| | 4' 6" | 5' 0" | 5' 6" | 6' 0" | 6" 6" | 7' 0" | 7' 6" | 8' 0" |
| 7/8" | .90 | .98 | 1.10 | 1.15 | 1.23 | 1.30 | 1.35 | 1.40 |
| 1" | .93 | 1.00 | 1.15 | 1.25 | 1.28 | 1.38 | 1.40 | 1.54 |
| 1-1/4" | .95 | 1.10 | 1.25 | 1.28 | 1.30 | 1.40 | 1.45 | 1.50 |
| 1-1/2" | .98 | 1.15 | 1.28 | 1.33 | 1.38 | 1.43 | 1.48 | 1.58 |
| 1-3/4" | 1.00 | 1.25 | 1.33 | 1.40 | 1.43 | 1.50 | 1.55 | 1.60 |
| 2" | 1.10 | 1.28 | 1.40 | 1.45 | 1.50 | 1.58 | 1.60 | 1.65 |
| 2-1/4" | 1.15 | 1.33 | 1.45 | 1.48 | 1.55 | 1.60 | 1.63 | 1.68 |
| 2-1/2" | 1.25 | 1.40 | 1.48 | 1.53 | 1.60 | 1.63 | 1.68 | 1.70 |

Manhours are based on overall length of anchor bolt from end to end including hook and are average for all heights.

Manhours are for the installation of template and bolt and sleeve for the size and length as outlined above.

All bolts listed above are assumed to be sleeved.

When converting above manhours to labor dollars, consideration should be given to the type of crew or crafts involved in the installation of above anchor bolts and a crew composite rate applied accordingly.

Manhours do not include fabrication of bolts.

Manhours do not include field engineering time spent aligning and checking anchor bolts. This is usually a part of field overhead and should be considered as such.

For sizes not listed, take the next highest listing.

# INSTALLATION OF STRAIGHT TYPE ANCHOR BOLTS

MANHOURS EACH

| Size | Manhours Each for Overall Length | | | | | |
|---|---|---|---|---|---|---|
| | 0' 8" | 1' 0" | 1' 6" | 2' 0" | 2' 6" | 3' 0" |
| 1/4" | .10 | .12 | .15 | .18 | .20 | .25 |
| 3/8" | .10 | .12 | .18 | .20 | .22 | .28 |
| 1/2" | .10 | .15 | .18 | .20 | .25 | .30 |
| 5/8" | .15 | .15 | .20 | .22 | .28 | .33 |
| 3/4" | .15 | .18 | .20 | .25 | .30 | .38 |
| 1" | .15 | .18 | .23 | .25 | .33 | .40 |

# LOOPS AND SCREW ANCHORS

MANHOURS EACH

| Item | Manhours |
|---|---|
| Coil Loops–½"x4" through 1½"x12" | 0.15 |
| Screw anchors–½" through 1½" | 0.12 |

Manhours are based on the installation of template and bolt for the size and length as outlined above and are average for all heights.

If a mixed crew of various crafts is used in the setting of above bolts, consideration should be given this when arriving at a composite rate for the conversion of manhours to labor dollars.

Manhours do not include engineering time spent in the aligning or checking of bolts. This is usually a part of field overhead and should be considered as such.

For sizes not listed, take the next highest listing.

# SHEET METAL BOX-OUT FORMS & SLEEVES

MANHOURS REQUIRED EACH

| ITEM | MANHOURS | | | |
|---|---|---|---|---|
| | Carpenter | Laborer | Truck Driver | Total |
| Box-Out Forms on Deck | | | | |
| Sizes 2" x 6" through 12" x 20" | .334 | .250 | .100 | .684 |
| Sleeves | | | | |
| Sizes 1-1/2" through 24" Round | | | | |
| Set on deck forms | .334 | .084 | .050 | .468 |
| Set on hand set wall forms | .500 | .100 | .050 | .650 |
| Set on gang wall forms | .667 | .167 | .050 | .884 |
| Set on beam forms | .500 | .084 | .050 | .634 |

Manhours include checking out of job storage, handling, hauling, and installing in position as outlined above.

Manhours do not include placement of forms or other concrete items. See respective tables for these time frames.

# REINFORCING RODS AND MESH

MANHOURS PER UNITS LISTED

| Item | Manhours | |
|---|---|---|
| | Per Ton | Per cwt |
| Unload, Sort & Pile Rods | 1.75 | 0.0875 |
| Fabricate – Cut and Bend | | |
| 1/2" round and larger | 6.00 | 0.3000 |
| 3/8" round and smaller | 11.48 | 0.5740 |
| Place loose without tieing | | |
| 3/4" round and larger | 7.25 | 0.3625 |
| 5/8" round and smaller | 8.78 | 0.4390 |
| Place and Tie Rods | | |
| Walls, columns, etc. | 16.50 | 0.8250 |
| Floors | 22.10 | 1.1050 |
| Average All Operations – All Sizes | | |
| Without tieing | 18.51 | 0.9250 |
| With tieing | 29.79 | 1.4900 |

| Item | Manhours per 100 sq ft | |
|---|---|---|
| Welded Wire Mesh | | |
| Cut & Place | .80 | |

It is the intent of the above listed manhours to cover necessary time for the various operations as listed above for the handling, fabrication and placement of reinforcing steel and mesh.

Manhours do not include the pouring of concrete or the placement of other concrete items. See respective tables for these charges.

The manhours cover necessary time for the various operations as listed above and as outlined in the introduction to this section.

# CONCRETE FOR MANHOLES & SEPTIC TANKS

MANHOURS PER CUBIC YARD

| Item | MANHOURS | | | | |
|---|---|---|---|---|---|
| | Laborer | Carpenter | Oper. Engr. | Oiler | Total |
| Square Type | | | | | |
| Slab on Ground | | | | | |
| Direct from truck | .53 | .04 | – | – | .57 |
| Chut e | .70 | – | – | – | .70 |
| Crane & Bucket | .88 | – | .03 | .03 | .94 |
| Walls | | | | | |
| Direct from truck | .55 | .04 | – | – | .59 |
| Chute | 1.00 | – | – | – | 1.00 |
| Crane & bucket | .99 | – | .04 | .04 | 1.07 |
| Elevated Slabs | | | | | |
| Ramp & buggies | 1.56 | – | .16 | – | 1.72 |
| Crane & Bucket | 1.09 | – | .10 | .10 | 1.29 |

Manhours are for placement and vibration of concrete for items outlined above.

Manhours do not include fabrication or erection of formwork, the placement of reinforcing steel or the finishing of concrete. See respective tables for these charges.

# CONCRETE FOR BOX TYPE CULVERTS AND THRUST BLOCKS

MANHOURS PER CUBIC YARD

| Item | MANHOURS | | | |
|---|---|---|---|---|
| | Laborer | Oper. Engr. | Oiler | Total |
| Box Type Culverts | | | | |
| Slab on Ground | .88 | .03 | .03 | .94 |
| Walls | .99 | .04 | .04 | 1.07 |
| Wing Walls | 1.00 | .05 | .05 | 1.10 |
| Elevated | 1.09 | .10 | .10 | 1.29 |
| Thrust Blocks | 1.00 | .05 | .05 | 1.10 |

Manhours are for the placement and vibration of concrete for items outlined above.

Manhours do not include the fabrication or erection of forms, the placement of reinforcing steel or the finishing of concrete. See respective tables for these charges.

# CONCRETE FOR EQUIPMENT FOUNDATIONS

MANHOURS PER CUBIC YARD

| Item | Manhours | | | |
|---|---|---|---|---|
| | Laborer | Oper. Engr. | Oiler | Total |
| **Square Pads** | | | | |
| Crane & bucket | 1.50 | .19 | .19 | 1.88 |
| Crane, bucket & buggies | 2.00 | .25 | .25 | 2.50 |
| **Offset, Skewed & Angled** | | | | |
| Crane & bucket | 2.44 | .38 | .38 | 3.20 |
| Crane, bucket & buggies | 3.25 | .50 | .50 | 4.25 |

Manhours are for the placement and vibration of concrete for the above items, in accordance with the introduction to this section.

Square Pad manhours are based on pouring of square pads to four feet high either integral with floor or over pre-set dowels on pre-poured floor.

Offset, skewed or angled manhours are based on that of pouring a large and bulky foundation with offsets or angles or both.

Manhours do not include finishing operations. See respective tables for these charges.

# CONCRETE FINISH

MANHOURS PER SQUARE FOOT

| Surface Finish | Manhours | | |
|---|---|---|---|
| | Cement Finisher | Laborer | Total |
| Carborundum Rub | .045 | – | .045 |
| Remove Fins or Ties - Point & Patch | .030 | – | .030 |
| Machine Trowel & Hand Burnish | .015 | – | .015 |
| Hand Steel Trowel | .030 | – | .030 |
| Woodfloat | .001 | – | .001 |
| Broom | .003 | – | .003 |
| Screeding Off | .003 | .003 | .006 |
| Cure & Protect | .002 | – | .002 |

**Manhours are for the above types of finish and include all necessary operations as may be required.**

**Manhours do not include the placement of concrete or concrete items. See respective tables for these charges.**

**In most areas, craft jurisdiction prevents laborers from helping cement finishers. Should this be the case, use total hours as listed above for cement finisher hours.**

# BRICK AND BLOCK MANHOLES AND PLASTER

MANHOURS PER UNITS LISTED

| Item | Unit | Mason | Plasterer | Hod Carrier | Total |
|---|---|---|---|---|---|
| Common Brick - tapered radial walls | 100 ea. | 1.4 | – | 1.2 | 2.6 |
| Concrete Brick - Tapered radial walls | 100 ea. | 1.6 | – | 1.2 | 2.8 |
| Concrete Block - square walls | 100 sq. ft. | 5.3 | – | 5.0 | 10.3 |
| Plaster walls | Sq. yd. | – | 0.1 | 0.9 | 1.0 |

Above manhours include mixing mortar and handling and placing masonry units.

Manhours do not include excavation or placement of concrete items not outlined above. See respective tables for these charges.

# Section 10

# MISCELLANEOUS STEEL, FRAMING, AND SUPPORTS

This section includes manhour tables for the fabrication and erection of various miscellaneous steel, framing, and supports that the mechanical contractor may provide.

In many cases the drawings will not show supports for a particular piece of equipment or ductwork, but the specifications will state that they are to be furnished by the contractor. Thus, bidding the job becomes the estimator's problem. Some of the following tables show average weights for various types of such supports as well as average manhours required to fabricate and erect them.

These manhour tables are based on averages of many projects under varied conditions where strict methods and preplanning were followed and strict reporting of time spent was recorded in accordance with the notes that appear with each table.

The manhours include time allowance to complete all necessary labor for the particular operation as outlined in each table.

# MISCELLANEOUS STEEL AND IRON

MANHOURS PER UNITS LISTED

| Item | Unit | Iron Worker Manhours |
|---|---|---|
| Embedded angles | | |
| Unload | cwt | .25 |
| Fabricate | cwt | 2.25 |
| Install | cwt | 1.75 |
| Louver framing | | |
| Unload | cwt | 0.25 |
| Fabricate | cwt | 2.25 |
| Install | cwt | 2.00 |
| Duct framing | | |
| Fabricate | cwt | 2.25 |
| Erect | cwt | 2.50 |
| Trench Covers | | |
| Install Plate | 100 sq. ft. | 10.00 |
| Install grating | 100 sq. ft. | 14.00 |
| Pipe sleeves | | |
| Fabricate | cwt | 3.00 |
| Install | cwt | 2.75 |

Manhours include handling, hauling, fabricating and erecting miscellaneous steel items as outlined above and as related to this type of work.

Manhours do not include sheet metal portion of ducts or scaffolding, See respective tables for these charges.

# SUPPORTS FOR FAN AND MOTOR UNITS

NET MANHOURS PER UNIT

| Unit Weight of Fans & Motors | Steel Hangers & Supports in Pounds | MANHOURS | | |
|---|---|---|---|---|
| | | Fabrication | Erection | Total |
| 400 | 150 | 3.9 | 3.1 | 7.0 |
| 500 | 150 | 3.9 | 3.1 | 7.0 |
| 600 | 150 | 3.9 | 3.1 | 7.0 |
| 700 | 200 | 4.6 | 3.8 | 8.4 |
| 1300 | 300 | 7.7 | 6.3 | 14.0 |
| 1800 | 375 | 9.6 | 7.9 | 17.5 |
| 2500 | 375 | 9.6 | 7.9 | 17.5 |
| 3900 | 375 | 9.6 | 7.9 | 17.5 |
| 5000 | 500 | 11.7 | 9.0 | 20.7 |
| 6000 | 500 | 11.7 | 9.0 | 20.7 |

Manhours include handling, hauling, fabricating and installing supports for fans and motors outlined above.

Manhours do not include scaffolding or installation of fans and motors. See respective tables for these charges.

# SUPPORTS FOR HEATING AND VENTILATING UNITS

NET MANHOURS PER UNIT

| Weight of Unit in Pounds | Steel Hangers & Supports in Pounds | MANHOURS | | |
|---|---|---|---|---|
| | | Fabrication | Erection | Total |
| 300 | 100 | 3.0 | 2.6 | 5.6 |
| 450 | 150 | 3.9 | 3.1 | 7.0 |
| 500 | 200 | 4.6 | 3.8 | 8.4 |
| 600 | 300 | 7.7 | 6.3 | 14.0 |
| 900 | 400 | 9.6 | 7.9 | 17.5 |
| 1500 | 600 | 13.7 | 9.1 | 22.8 |
| 1600 | 650 | 14.8 | 9.8 | 24.6 |
| 2500 | 650 | 14.8 | 9.8 | 24.6 |
| 2600 | 700 | 15.8 | 10.6 | 26.4 |
| 3500 | 700 | 15.8 | 10.6 | 26.4 |
| 3700 | 750 | 16.9 | 11.3 | 28.2 |
| 4300 | 800 | 18.0 | 12.0 | 30.0 |
| 4500 | 800 | 18.0 | 12.0 | 30.0 |

Manhours include handling, hauling fabricating and installing supports for heating and ventilating units as outlined above.

Manhours do not include scaffolding or installation of heating and ventilating units. See respective tables for these charges.

# SUPPORTS FOR SELF-CONTAINED AIR-CONDITIONING UNITS

NET MANHOURS PER UNIT

| Refrigeration Tons | Air Conditioning Unit in Pounds | Steel Hangers & Supports in Pounds | MANHOURS | | |
|---|---|---|---|---|---|
| | | | Fabrication | Erection | Total |
| 6 | 2000 | 650 | 13.5 | 11.1 | 24.6 |
| 10 | 3000 | 700 | 14.5 | 11.9 | 26.4 |
| 15 | 3800 | 750 | 15.5 | 12.7 | 28.2 |
| 20 | 4000 | 800 | 16.5 | 13.5 | 30.0 |
| 30 | 4500 | 900 | 18.8 | 15.4 | 34.2 |
| 40 | 6000 | 1350 | 30.6 | 20.4 | 51.0 |
| 50 | 8000 | 1700 | 38.2 | 25.4 | 63.6 |

**Manhours include handling, hauling, fabricating and installing supports for air-conditioning units as outlined above.**

**Manhours do not include scaffolding or installation of air-conditioning units. See respective tables for these charges.**

# SUPPORTS FOR AIR HANDLING UNITS

MANHOURS PER UNIT

| Capacity CFM | Steel Hangers and Supports Pounds | MANHOURS | | |
|---|---|---|---|---|
| | | Fabrication | Erection | Total |
| Single Zone Units | | | | |
| 1,000 | 500 | 7.9 | 6.5 | 14.4 |
| 2,500 | 600 | 9.9 | 8.1 | 18.0 |
| 6,000 | 625 | 13.2 | 10.8 | 24.0 |
| 14,000 | 750 | 28.8 | 19.2 | 48.0 |
| 24,000 | 800 | 33.8 | 22.6 | 56.4 |
| 30,000 | 900 | 39.6 | 26.4 | 66.0 |
| Multi-Zone Units | | | | |
| 4,000 | 550 | 8.4 | 7.3 | 15.7 |
| 6,000 | 625 | 13.2 | 10.8 | 24.0 |
| 10,000 | 650 | 15.8 | 13.0 | 28.8 |
| 15,000 | 775 | 31.3 | 21.4 | 52.7 |
| 22,000 | 800 | 33.8 | 22.6 | 56.4 |
| 30,000 | 900 | 39.6 | 26.4 | 66.0 |

Manhours include handling, hauling, fabricating, and installing supports for air handling units as outlined above.

Manhours do not include scaffolding or installation of air handling units. See respective tables for these charges.

# Section 11

# TECHNICAL INFORMATION

This manual is solely intended for the estimation of labor and not for the design of systems or items. Therefore, this section has been held to a minimum and includes only information that will benefit the estimator in the preparation of his estimate.

This section contains a table showing the conversion of minutes to decimal hours, a manhour table for the installation of patent scaffolding, a table showing the percentage comparison of other metals to mild steel, and many weight tables showing the weights of various metals, pipe, and fittings for estimating freight costs.

# MINUTES TO DECIMAL HOURS CONVERSION TABLE

| Minutes | Hours | Minutes | Hours |
|---|---|---|---|
| 1 | .017 | 31 | .517 |
| 2 | .034 | 32 | .534 |
| 3 | .050 | 33 | .550 |
| 4 | .067 | 34 | .567 |
| 5 | .084 | 35 | .584 |
| 6 | .100 | 36 | .600 |
| 7 | .117 | 37 | .617 |
| 8 | .135 | 38 | .634 |
| 9 | .150 | 39 | .650 |
| 10 | .167 | 40 | .667 |
| 11 | .184 | 41 | .684 |
| 12 | .200 | 42 | .700 |
| 13 | .217 | 43 | .717 |
| 14 | .232 | 44 | .734 |
| 15 | .250 | 45 | .750 |
| 16 | .267 | 46 | .767 |
| 17 | .284 | 47 | .784 |
| 18 | .300 | 48 | .800 |
| 19 | .317 | 49 | .817 |
| 20 | .334 | 50 | .834 |
| 21 | .350 | 51 | .850 |
| 22 | .368 | 52 | .867 |
| 23 | .384 | 53 | .884 |
| 24 | .400 | 54 | .900 |
| 25 | .417 | 55 | .917 |
| 26 | .434 | 56 | .934 |
| 27 | .450 | 57 | .950 |
| 28 | .467 | 58 | .967 |
| 29 | .484 | 59 | .984 |
| 30 | .500 | 60 | 1.000 |

# ERECT AND DISMANTLE PATENT SCAFFOLDING

DIRECT LABOR - MANHOURS PER SECTION

| Length | 1 or 2 Sections High | | | More than 2 Sections High | | |
|---|---|---|---|---|---|---|
| | Erect | Dismtl | Total | Erect | Dismtl | Total |
| One to Two sections long | 1.40 | 1.00 | 2.40 | 1.70 | 1.20 | 2.90 |
| Three to five sections long | .90 | .60 | 1.50 | 1.00 | .70 | 1.70 |
| Six or more sections long | .70 | .40 | 1.10 | .90 | .50 | 1.40 |

Manhours are for installation of patent tubular scaffolding with 2″ plank topping. Sections are 7′ long by 5′ wide by 5′ high. Manhours include transporting scaffolding and materials from storage, erection, leveling and securing scaffolding, installation of 2″ planking, dismantling of scaffolding and transporting scaffolding and materials back to storage.

Manhours are for patent type tubular scaffolding only and are not intended to suffice for homemade type scaffolding.

# COMPARISON PERCENTAGE OTHER METALS TO MILD STEEL

PERCENTAGES

| | PERCENTAGE FACTORS | |
|---|---|---|
| Materials | Burn or Weld | Hacksaw, Grind or File |
| Cast Iron | 2.00 | 1.75 |
| Cast Steel | 1.90 | 1.25 |
| Stainless Steel | 1.75 | 3.40 |
| Nickel | 1.40 | 3.25 |
| Monel | 1.75 | 3.25 |
| Copper | 1.15 | .50 |
| Cast Brass | 1.25 | .75 |
| Brass | 1.20 | .45 |
| Bronze | 1.25 | .75 |
| Aluminum | .60 | .40 |

# WEIGHT TABLE—GALVANIZED STEEL SHEET

| Gauge | Pounds Per Square Foot |
|---|---|
| 10 | 5.781 |
| 11 | 5.156 |
| 12 | 4.531 |
| 14 | 3.281 |
| 16 | 2.656 |
| 18 | 2.156 |
| 20 | 1.656 |
| 22 | 1.406 |
| 24 | 1.156 |
| 26 | .906 |
| 27 | .844 |
| 28 | .781 |
| 30 | .656 |

# WEIGHT TABLE—ALUMINUM SHEET & PLATE

| B & S Gauge Number | Thickness (In Inches) | Approximate Weight Per Square Foot | |
|---|---|---|---|
| | | Sheet | Plate |
| – | * .190 | 2.68 | – |
| – | .188 | 2.65 | – |
| – | * .160 | 2.26 | – |
| – | .156 | 2.20 | – |
| – | .125 | 1.76 | – |
| 10 | .102 | 1.44 | – |
| – | * .100 | 1.41 | – |
| 11 | .091 | 1.28 | – |
| – | * .090 | 1.27 | – |
| 12 | .081 | 1.14 | – |
| – | * .080 | 1.13 | – |
| 13 | .072 | 1.04 | – |
| – | * .071 | 1.00 | – |
| 14 | .064 | .903 | – |
| – | * .063 | .889 | – |
| 16 | .051 | .716 | – |
| – | * .050 | .706 | – |
| 18 | .040 | .568 | – |
| 20 | .032 | .450 | – |
| 22 | .025 | .357 | – |
| 24 | .020 | .283 | – |
| 26 | .016 | .225 | – |
| 28 | .012 | .178 | – |
| 30 | .010 | .141 | – |
| 32 | .008 | .113 | – |
| 34 | .006 | .085 | – |
| – | 2.000 | – | 28.2 |
| – | 1.750 | – | 24.7 |
| – | 1.500 | – | 21.2 |
| – | 1.250 | – | 17.6 |
| – | 1.000 | – | 14.1 |
| – | .875 | – | 12.3 |
| – | .750 | – | 10.6 |
| – | .625 | – | 8.8 |
| – | .500 | – | 7.1 |
| – | .375 | – | 5.28 |
| – | .313 | – | 4.40 |
| – | .250 | – | 3.52 |

* American Standard Preferred Thickness.

# WEIGHT TABLE—BRASS SHEET

| Thickness in Inches | B & S Gauge | Pounds Per Square Foot | Thickness in Inches | B & S Gauge | Pounds Per Square Foot |
|---|---|---|---|---|---|
| 1.000 | – | 44.06 | .0571 | 15 | 2.516 |
| .875 | – | 38.56 | .0508 | 16 | 2.238 |
| .750 | – | 33.05 | .0453 | 17 | 1.996 |
| .625 | – | 27.54 | .0403 | 18 | 1.776 |
| .500 | – | 22.03 | .0359 | 19 | 1.582 |
| .4600 | 4/0 | 20.27 | .0320 | 20 | 1.410 |
| .4096 | 3/0 | 18.05 | .0285 | 21 | 1.256 |
| .375 | – | 16.52 | .0253 | 22 | 1.115 |
| .3648 | 2/0 | 16.07 | .0226 | 23 | .9958 |
| .3249 | 1/0 | 14.32 | .0201 | 24 | .8857 |
| .3125 | – | 13.77 | .0179 | 25 | .7887 |
| .2893 | 1 | 12.75 | .0159 | 26 | .7006 |
| .2576 | 2 | 11.35 | .0142 | 27 | .6257 |
| .250 | – | 11.02 | .0126 | 28 | .5552 |
| .2294 | 3 | 10.11 | .0113 | 29 | .4979 |
| .2043 | 4 | 9.002 | .0100 | 30 | .4406 |
| .1875 | – | 8.262 | .0089 | 31 | .3922 |
| .1819 | 5 | 8.015 | .0080 | 32 | .3528 |
| .1620 | 6 | 7.138 | .0071 | 33 | .3129 |
| .1443 | 7 | 6.358 | .0063 | 34 | .2776 |
| .1285 | 8 | 5.662 | .0056 | 35 | .2468 |
| .125 | – | 5.508 | .0050 | 36 | .2203 |
| .1144 | 9 | 5.041 | .0045 | 37 | .1983 |
| .1019 | 10 | 4.490 | .0040 | 38 | .1763 |
| .0907 | 11 | 3.997 | .0035 | 39 | .1542 |
| .0808 | 12 | 3.560 | .0031 | 40 | .1366 |
| .0720 | 13 | 3.173 | .0028 | 41 | .1234 |
| .0641 | 14 | 2.825 | .0025 | 42 | .1101 |

# WEIGHT TABLE—COPPER SHEET

| Thickness In Inches | Nearest B & S Gauge | Pounds Per Square Foot | Thickness In Inches | Nearest B & S Gauge | Pounds Per Square Foot |
|---|---|---|---|---|---|
| .3451 | 2/0 | 16 | .0755 | 13 | 3.5 |
| .3235 | 1/0 | 15 | .0647 | 14 | 3 |
| .3019 | 1 | 14 | .0593 | 15 | 2.75 |
| .2804 | 1 | 13 | .0539 | 16 | 2.5 |
| .2588 | 2 | 12 | .0485 | 16 | 2.25 |
| .2372 | 3 | 11 | .0431 | 17 | 2 |
| .2157 | 4 | 10 | .0377 | 19 | 1.75 |
| .2049 | 4 | 9.5 | .0323 | 20 | 1.5 |
| .1941 | 4 | 9 | .0270 | 21 | 1.25 |
| .1833 | 5 | 8.5 | .0243 | 22 | 1.13 |
| .1725 | 5 | 8 | .0216 | 23 | 1 |
| .1617 | 6 | 7.5 | .0189 | 25 | 0.87 |
| .1510 | 7 | 7 | .0162 | 26 | 0.75 |
| .1402 | 7 | 6.5 | .0135 | 27 | 0.63 |
| .1294 | 8 | 6 | .0108 | 29 | 0.5 |
| .1186 | 9 | 5.5 | .0081 | 32 | 0.37 |
| .1078 | 10 | 5 | .0054 | 35 | 0.25 |
| .0970 | 10 | 4.5 | .0027 | 41 | 0.13 |
| .0863 | 11 | 4 | | | |

# WEIGHT TABLE—STAINLESS STEEL SHEET

| Thickness In Inches | U.S. Std. Gauge | Pounds Per Square Foot |
|---|---|---|
| .1406 | 10 | 5.906 |
| .125 | 11 | 5.25 |
| .1093 | 12 | 4.593 |
| .0937 | 13 | 3.937 |
| .0781 | 14 | 3.281 |
| .0625 | 16 | 2.625 |
| .050 | 18 | 2.10 |
| .040 | – | 1.648 |
| .0375 | 20 | 1.575 |
| .035 | – | 1.442 |
| .0312 | 22 | 1.3125 |
| .025 | 24 | 1.05 |
| .020 | – | .824 |
| .0187 | 26 | .7875 |
| .016 | – | .659 |

# WEIGHT TABLE—SEAMLESS STEEL PIPE

POUNDS PER FOOT

| Pipe Size Inches | STANDARD WEIGHT | | EXTRA STRONG | |
|---|---|---|---|---|
| | Wall Inches | Pounds Per Foot | Wall Inches | Pounds Per Foot |
| 2 | .154 | 3.65 | .218 | 5.02 |
| 2-1/2 | .203 | 5.79 | .276 | 7.66 |
| 3 | .216 | 7.57 | .300 | 10.25 |
| 3-1/2 | .226 | 9.10 | .318 | 12.50 |
| 4 | .237 | 10.79 | .337 | 14.98 |
| 5 | .258 | 14.61 | .375 | 20.77 |
| 6 | .280 | 18.97 | .432 | 28.57 |
| 8 | .277 | 24.69 | .500 | 43.38 |
| 8 | .322 | 28.55 | – | – |
| 10 | .279 | 31.20 | .500 | 54.74 |
| 10 | .365 | 40.48 | .593 | 64.33 |
| 12 | .330 | 43.77 | .500 | 65.42 |
| 12 | .375 | 49.56 | .687 | 88.51 |

# WEIGHT TABLE—BUTTWELD PIPE

POUNDS PER FOOT

| Pipe Size Inches | STANDARD WEIGHT | | EXTRA STRONG | |
|---|---|---|---|---|
| | Wall Inches | Pounds Per foot | Wall Inches | Pounds Per Foot |
| 1/8 | .068 | .24 | .095 | .31 |
| 1/4 | .088 | .42 | .119 | .54 |
| 3/8 | .091 | .56 | .126 | .74 |
| 1/2 | .109 | .85 | .147 | 1.09 |
| 3/4 | .113 | 1.13 | .154 | 1.47 |
| 1 | .133 | 1.68 | .179 | 2.17 |
| 1-1/4 | .140 | 2.28 | .191 | 3.00 |
| 1-1/2 | .145 | 2.73 | .200 | 3.63 |
| 2 | .154 | 3.68 | .218 | 5.02 |
| 2-1/2 | .203 | 5.82 | .276 | 7.66 |
| 3 | .216 | 7.62 | .300 | 10.25 |
| 3-1/2 | — | — | .318 | 12.51 |
| 4 | .237 | 10.89 | .337 | 14.98 |

# WEIGHT TABLE—COPPER PIPE

POUNDS PER FOOT

| Pipe Size Inches | TYPE K | | TYPE L | | TYPE M | |
|---|---|---|---|---|---|---|
| | Wall Inches | Pounds Per Foot | Wall Inches | Pounds Per Ft. | Wall Inches | Pounds Per Foot |
| 3/8 | .049 | .269 | .035 | .198 | .025 | .145 |
| 1/2 | .049 | .344 | .040 | .285 | .028 | .204 |
| 5/8 | .049 | .418 | .042 | .362 | – | – |
| 3/4 | .065 | .641 | .045 | .455 | .032 | .328 |
| 1 | .065 | .839 | .050 | .655 | .035 | .465 |
| 1-1/4 | .065 | 1.04 | .055 | .884 | .042 | .681 |
| 1-1/2 | .072 | 1.36 | .060 | 1.14 | .049 | .940 |
| 2 | .083 | 2.06 | .070 | 1.75 | .058 | 1.46 |
| 2-1/2 | .095 | 2.93 | .080 | 2.48 | .065 | 2.03 |
| 3 | .109 | 4.00 | .090 | 3.33 | .072 | 2.68 |
| 3-1/2 | .120 | 5.12 | .100 | 4.29 | .083 | 3.58 |
| 4 | .134 | 6.51 | .110 | 5.38 | .095 | 4.66 |
| 5 | .160 | 9.67 | .125 | 7.61 | .109 | 6.66 |
| 6 | .192 | 13.87 | .140 | 10.20 | .122 | 8.91 |
| 8 | .271 | 25.90 | .200 | 19.30 | .170 | 16.46 |
| 10 | .338 | 40.30 | .250 | 30.10 | .212 | 25.60 |
| 12 | .405 | 57.80 | .280 | 40.40 | .254 | 36.70 |

# WEIGHT TABLE—POLYVINYL CHLORIDE PIPE

POUNDS PER FOOT

| Nominal Size Inches | OD Size Inches | Pounds Per Foot | |
|---|---|---|---|
| | | Schedule 40 | Schedule 80 |
| 1/2 | 0.84 | 0.15 | 0.19 |
| 3/4 | 1.05 | 0.20 | 0.26 |
| 1 | 1.31 | 0.30 | 0.38 |
| 1-1/4 | 1.66 | 0.41 | 0.55 |
| 1-1/2 | 1.90 | 0.49 | 0.65 |
| 2 | 2.37 | 0.64 | 0.91 |
| 2-1/2 | 2.87 | 1.10 | 1.50 |
| 3 | 3.50 | 1.35 | 2.00 |
| 4 | 4.50 | 1.95 | 3.00 |
| 6 | 6.25 | 3.50 | 5.85 |
| 8 | 8.25 | 5.75 | 8.00 |
| 10 | 10.25 | 8.00 | 10.75 |
| 12 | 12.25 | 10.00 | 13.25 |

# 150-POUND CAST-IRON PIPE

PIPE POUNDS PER LINEAR FOOT
LEAD AND JUTE POUNDS PER JOINT

| Pipe Size Inches | Pipe Pounds Per Foot | Pounds Per Joint | |
|---|---|---|---|
| | | Lead | Jute |
| 3 | 12.5 | 6.85 | 0.20 |
| 4 | 16.5 | 9.00 | 0.25 |
| 6 | 26.0 | 12.00 | 0.38 |
| 8 | 37.0 | 15.55 | 0.50 |
| 10 | 49.2 | 18.25 | 0.57 |
| 12 | 64.0 | 21.50 | 0.60 |
| 14 | 80.7 | 25.00 | 0.82 |
| 16 | 97.5 | 34.00 | 0.95 |
| 18 | 117.0 | 38.00 | 1.15 |

# WEIGHT TABLE—STANDARD WALL STEAMLESS STEEL WELDING FITTINGS

POUNDS EACH

| Pipe Size Inches | LONG RADIUS | | | SHORT RADIUS | | Reducing Ells |
|---|---|---|---|---|---|---|
| | 90° Ells | 45° Ells | 180° Returns | 90° Ells | 180° Returns | |
| 1/2 | .25 | .25 | .25 | – | – | – |
| 3/4 | .25 | .25 | .50 | – | – | – |
| 1 | .50 | .25 | 1. | .25 | .50 | – |
| 1-1/4 | .50 | .50 | 1. | .50 | 1. | – |
| 1-1/2 | .75 | .50 | 2. | .50 | 1. | – |
| 2 | 2. | 1. | 3. | 1. | 2. | 1. |
| 2-1/2 | 3. | 2. | 6. | 2. | 4. | 2. |
| 3 | 5. | 3. | 10. | 3. | 6. | 4. |
| 3-1/2 | 7. | 4. | 13. | 4. | 9. | 5. |
| 4 | 9. | 4. | 18. | 6. | 12. | 8. |
| 5 | 16. | 8. | 30. | 10. | 19. | 12. |
| 6 | 24. | 12. | 50. | 18. | 35. | 18. |
| 8 | 50. | 23. | 95. | 34. | 68. | 35. |
| 10 | 88. | 43. | 177. | 58. | 115. | 61. |
| 12 | 125. | 62. | 230. | 80. | 155. | 96. |
| 14 | 160. | 80. | 325. | 105. | 210. | – |
| 16 | 206. | 100. | 412. | 132. | 260. | – |
| 18 | 260. | 126. | 510. | 167. | 330. | – |
| 20 | 320. | 160. | 640. | 210. | 410. | – |
| 24 | 460. | 238. | 890. | 298. | 590. | – |

# WEIGHT TABLE—STANDARD WALL SEAMLESS STEEL WELDING FITTINGS

POUNDS EACH

| Pipe Size Inches | WELDING TEES | | Lap Joint Stud Ends | CROSSES | | Caps |
|---|---|---|---|---|---|---|
| | Full Size | Reducing | | Full | Reducing | |
| 3/4 | .50 | .50 | .25 | – | – | .25 |
| 1 | 1. | 1. | .75 | – | – | .25 |
| 1-1/4 | 1. | 2. | 1. | 2. | 2. | .25 |
| 1-1/2 | 2. | 2. | 1. | 2. | 2. | .50 |
| 2 | 4. | 3. | 2. | 3. | 2. | .50 |
| 2-1/2 | 6. | 5. | 4. | 5. | 4. | 1. |
| 3 | 7. | 6. | 5. | 8. | 8. | 2. |
| 3-1/2 | 9. | 8. | 6. | 11. | 10. | 2. |
| 4 | 12. | 11. | 7. | 14. | 13. | 3. |
| 5 | 21. | 20. | 12. | 28. | 24. | 4. |
| 6 | 34. | 32. | 16. | 43. | 40. | 6. |
| 8 | 55. | 52. | 24. | 67. | 63. | 12. |
| 10 | 85. | 81. | 36. | 103. | 99. | 20. |
| 12 | 120. | 114. | 47. | 145. | 138. | 30. |
| 14 | 165. | 155. | 60. | 198. | 192. | 36. |
| 16 | 195. | 186. | 70. | 244. | 229. | 40. |
| 18 | 249. | 230. | 94. | 299. | 283. | 54. |
| 20 | 342. | 336. | 105. | 414. | 405. | 75. |
| 24 | 528. | 516. | 126. | 636. | 624. | 96. |

# WEIGHT TABLE—EXTRA STRONG SEAMLESS STEEL WELDING FITTINGS

POUNDS EACH

| Pipe Size Inches | LONG RADIUS | | | SHORT RADIUS | | |
|---|---|---|---|---|---|---|
| | $90^0$ Ells | $45^0$ Ells | $180^0$ Returns | $90^0$ Ells | $180^0$ Returns | Reducing Ells |
| 1/2 | .25 | .25 | .50 | – | – | – |
| 3/4 | .25 | .25 | .50 | – | – | – |
| 1 | .50 | .25 | 1 | – | – | – |
| 1-1/4 | 1. | .50 | 2. | – | – | – |
| 1-1/2 | 1. | .50 | 2. | .75 | 2. | – |
| 2 | 2. | 1. | 4. | 2. | 3. | 2. |
| 2-1/2 | 4. | 2. | 8. | 3. | 6. | 3. |
| 3 | 6. | 4. | 13. | 4. | 8. | 4. |
| 3-1/2 | 8. | 4. | 17. | 6. | 12. | 7. |
| 4 | 14. | 6. | 25. | 8. | 17. | 10. |
| 5 | 22. | 11. | 44. | 14. | 28. | 16. |
| 6 | 35. | 18. | 70. | 23. | 46. | 27. |
| 8 | 71. | 35. | 142. | 48. | 100. | 53. |
| 10 | 107. | 53. | 215. | 70. | 140. | 86. |
| 12 | 160. | 84. | 320. | 104. | 218. | 135. |
| 14 | 205. | 100. | 400. | 140. | 275. | – |
| 16 | 276. | 135. | 550. | 174. | 340. | – |
| 18 | 340. | 167. | 690. | 219. | 430. | – |
| 20 | 420. | 206. | 830. | 275. | 550. | – |
| 24 | 600. | 300. | 1200. | 392. | 780. | – |

# WEIGHT TABLE—STANDARD WALL SEAMLESS STEEL WELDING FITTINGS

POUNDS EACH

| Pipe Size Inches | WELDING TEES | | Lap Joint Stud Ends | CROSSES | | Caps |
|---|---|---|---|---|---|---|
| | Full Size | Reducing | | Full | Reducing | |
| 3/4 | .50 | .50 | .50 | — | — | .25 |
| 1 | 1. | 1. | 1. | — | — | .25 |
| 1-1/4 | 2. | 2. | 1. | 2. | 2. | .50 |
| 1-1/2 | 2. | 2. | 2. | 3. | 2. | .50 |
| 2 | 4. | 4. | 3. | 4. | 3. | .75 |
| 2-1/2 | 7. | 7. | 4. | 6. | 5. | 1. |
| 3 | 8. | 8. | 7. | 10. | 9. | 2. |
| 3-1/2 | 12. | 12. | 8. | 14. | 14. | 2. |
| 4 | 16. | 15. | 10. | 19. | 18. | 3. |
| 5 | 26. | 24. | 17. | 33. | 29. | 6. |
| 6 | 40. | 37. | 22. | 50. | 47. | 9. |
| 8 | 75. | 72. | 32. | 92. | 86. | 16. |
| 10 | 105. | 106. | 53. | 138. | 129. | 25. |
| 12 | 160. | 165. | 62. | 224. | 210. | 36. |
| 14 | 240. | 230 | 80. | 289. | 279. | 45. |
| 16 | 280. | 266. | 93. | 338. | 324. | 54. |
| 18 | 332. | 307. | 124. | 397. | 378. | 72. |
| 20 | 480. | 472. | 138. | 578. | 567. | 86. |
| 24 | 610. | 595. | 167. | 728. | 717. | 130. |

# WEIGHT TABLE—SEAMLESS STEEL CONCENTRIC & ECCENTRIC WELDING REDUCERS

POUNDS EACH

| Nominal Pipe Size Inches | WALL THICKNESS | |
|---|---|---|
| | Standard | Extra Strong |
| 1 x 3/8 – (1/2) – (3/4) | .50 | .50 |
| 1-1/4 x 1/2 – (3/4) – (1) | .50 | .50 |
| 1-1/2 x 1/2 – (3/4) | .50 | .75 |
| 1-1/2 x 1 – (1-1/4) | .75 | .75 |
| 2 x 3/4 – (1) – (1-1/4) – (1-1/2) | .75 | 1. |
| 2-1/2 x 1 – (1-1/4) – (1-1/2) | 1. | 2. |
| 2-1/2 x 2 | 2. | 2. |
| 3 x 1 – (1-1/4) – (1-1/2) | 2. | 2. |
| 3 x 2 – (2-1/2) | 2. | 3. |
| 3-1/2 x 1-1/4 – (1-1/2) | 2. | 3. |
| 3-1/2 x 2 – (2-1/2) – (3) | 3. | 4. |
| 4 x 1 – (1-1/4) – (1-1/2) | 3. | 4. |
| 4 x 2 – (2-1/2) – (3) – (3-1/2) | 3. | 4. |
| 5 x 2 – (2-1/2) – (3) – (3-1/2) | 5. | 7. |
| 5 x 4 | 6. | 8. |
| 6 x 2-1/2 – (3) – (3-1/2) | 8. | 10. |
| 6 x 4 – (5) | 8. | 12. |

# WEIGHT TABLE—SEAMLESS STEEL CONCENTRIC & ECCENTRIC WELDING REDUCERS

POUNDS EACH

| Nominal Pipe Size Inches | Wall Thickness | |
|---|---|---|
| | Standard | Extra Strong |
| 8 x 3 | 11. | 16. |
| 8 x (3-1/2)-4-5 | 12. | 17. |
| 8 x 6 | 13. | 19. |
| 10 x 4-5 | 20. | 26. |
| 10 x 6 | 21. | 30. |
| 10 x 8 | 22. | 30. |
| 12 x 4 | 30. | 38. |
| 12 x 6-8 | 32. | 40. |
| 12 x 10 | 34. | 44. |
| 14 x 6 | 58. | 78. |
| 14 x 8 | 59. | 79. |
| 14 x 10-12 | 60. | 79. |
| 16 x 8 | 68. | 88. |
| 16 x 10 | 70. | 89. |
| 16 x 12-14 | 71. | 90. |
| 18 x 10 | 82. | 112. |
| 18 x 12 | 83. | 113. |
| 18 x 14-16 | 84. | 114. |
| 20 x 12-14 | 120. | 167. |
| 20 x 16-18 | 124. | 169. |
| 24 x 16 | 145. | 190. |
| 24 x 18 | 148. | 195. |
| 24 x 20 | 150. | 200. |

# WEIGHT TABLE—FORGED STEEL FLANGES

POUNDS EACH

| Pipe | Welding Neck | | Slip-On | | Lap Joint | | Threaded | | Blind | |
|---|---|---|---|---|---|---|---|---|---|---|
| Size Inches | 150 Lb. | 300 Lb. | 150 Lb. | 300 Lb. | 150 Lb. | 300 Lb. | 150 Lb. | 300 Lb. | 150 Lb. | 300 Lb. |
| 1/2 | 2 | 2 | 1 | 2 | 1 | 2 | 1 | 2 | 2 | 2 |
| 3/4 | 2 | 3 | 2 | 2 | 2 | 2 | 2 | 2 | 2 | 3 |
| 1 | 3 | 4 | 2 | 3 | 2 | 3 | 2 | 3 | 2 | 4 |
| 1-1/4 | 3 | 5 | 2 | 4 | 2 | 4 | 2 | 4 | 3 | 6 |
| 1-1/2 | 4 | 7 | 3 | 6 | 3 | 6 | 3 | 6 | 3 | 7 |
| 2 | 6 | 8 | 5 | 7 | 5 | 7 | 5 | 7 | 4 | 8 |
| 2-1/2 | 10 | 12 | 8 | 10 | 8 | 10 | 8 | 10 | 7 | 12 |
| 3 | 11 | 18 | 9 | 13 | 9 | 14 | 10 | 14 | 9 | 16 |
| 3-1/2 | 12 | 20 | 11 | 16 | 11 | 16 | 12 | 16 | 13 | 21 |
| 4 | 16 | 26 | 13 | 24 | 12 | 24 | 13 | 24 | 17 | 28 |
| 5 | 21 | 36 | 15 | 29 | 13 | 26 | 15 | 31 | 20 | 37 |
| 6 | 26 | 45 | 17 | 36 | 18 | 38 | 20 | 36 | 27 | 48 |
| 8 | 42 | 69 | 28 | 56 | 28 | 55 | 30 | 56 | 47 | 79 |
| 10 | 54 | 100 | 40 | 77 | 36 | 88 | 41 | 80 | 67 | 122 |
| 12 | 88 | 142 | 61 | 113 | 60 | 139 | 65 | 110 | 123 | 183 |
| 14 | 114 | 206 | 83 | 159 | 77 | 184 | 85 | – | 139 | 241 |
| 16 | 142 | 249 | 106 | 210 | 104 | 234 | 93 | – | 187 | 315 |
| 18 | 165 | 306 | 109 | 253 | 146 | 305 | – | – | 217 | 414 |
| 20 | 197 | 369 | 148 | 307 | 159 | 375 | – | – | 283 | 515 |
| 24 | 268 | 519 | 204 | 490 | 195 | 530 | – | – | 415 | 800 |

# WEIGHT TABLE—FORGED STEEL FITTINGS

POUNDS EACH

| ASTMA – 105 | NORMAL PIPE SIZE – INCHES | | | | | | | | | | | |
|---|---|---|---|---|---|---|---|---|---|---|---|---|
| | 1/8 | 1/4 | 3/8 | 1/2 | 3/4 | 1 | 1-¼ | 1-½ | 2 | 2-½ | 3 | 4 |
| **2000 lb. Screwed Fittings:** | | | | | | | | | | | | |
| 90° Elbows | 1/4 | 1/4 | 1/4 | 1/2 | 3/4 | 1 | 2 | 2 | 4 | 6 | 10 | 23 |
| 45° Elbows | 1/8 | 1/8 | 1/4 | 1/2 | 1/2 | 1 | 1 | 2 | 3 | 8 | 12 | 19 |
| Tees | 1/4 | 1/4 | 1/4 | 1/2 | 1 | 1 | 2 | 3 | 5 | 9 | 13 | 27 |
| Crosses | 1/2 | 1/2 | 1/2 | 1 | 1 | 2 | 2 | 3 | 5 | 16 | 20 | 33 |
| Laterals | — | 1/4 | 1/2 | 1 | 2 | 2 | 3 | 4 | 7 | — | — | — |
| **3000 lb. Screwed Fittings:** | | | | | | | | | | | | |
| 90° Elbows | 1/4 | 1/4 | 1/2 | 1 | 1 | 2 | 3 | 4 | 5 | 11 | 15 | 30 |
| 45° Elbows | 1/4 | 1/4 | 1/2 | 3/4 | 1 | 2 | 2 | 3 | 4 | 7 | 10 | 19 |
| Tees | 1/4 | 1/2 | 1 | 1 | 2 | 2 | 3 | 5 | 7 | 13 | 20 | 40 |
| Crosses | 1/2 | 1/2 | 1 | 2 | 3 | 4 | 4 | 6 | 8 | 17 | 20 | 32 |
| Street Ells | — | 1/4 | 1/2 | 1/2 | 1 | 1 | 2 | 3 | 5 | — | — | — |
| Laterals | — | 1/2 | 1 | 2 | 3 | 5 | 6 | 11 | 14 | — | — | — |
| Couplings | 1/8 | 1/8 | 1/4 | 1/4 | 1/2 | 1 | 2 | 2 | 3 | 4 | 7 | 17 |
| Reducing Couplings | — | 1/8 | 1/4 | 1/4 | 1/2 | 1 | 2 | 2 | 3 | 4 | 7 | 17 |
| Half Couplings | 1/8 | 1/8 | 1/8 | 1/8 | 1/4 | 1/2 | 3/4 | 1 | 2 | 2 | 3 | 8 |
| Caps | 1/8 | 1/8 | 1/4 | 1/4 | 1/4 | 1/2 | 1 | 2 | 3 | 5 | 8 | 14 |
| **6000 lb. Screwed Fittings:** | | | | | | | | | | | | |
| 90° Elbows | 1/4 | 1/2 | 1 | 2 | 3 | 4 | 7 | 8 | 14 | 21 | 35 | — |
| 45° Elbows | 1/4 | 1/2 | 1/2 | 1 | 2 | 3 | 5 | 6 | 10 | 15 | 31 | — |
| Tees | 1/2 | 1 | 1 | 2 | 4 | 5 | 8 | 10 | 19 | 28 | 46 | — |
| Crosses | 1 | 1 | 2 | 3 | 4 | 6 | 11 | 12 | 22 | 28 | 54 | — |
| Street Ells | — | 1/4 | 1/2 | 1 | 2 | 2 | 4 | 6 | — | — | — | — |
| Laterals | — | — | 2 | 3 | 5 | 7 | 12 | 15 | — | — | — | — |
| Couplings | 1/8 | 1/4 | 1/4 | 1/2 | 1 | 2 | 2 | 4 | 8 | 11 | 14 | 24 |
| Reducing Couplings | — | 1/4 | 1/4 | 1/2 | 1 | 2 | 2 | 4 | 8 | 11 | 14 | 24 |
| Half Couplings | 1/8 | 1/8 | 1/8 | 1/4 | 1/2 | 1 | 1 | 2 | 4 | 5 | 7 | 12 |
| Caps | 1/8 | 1/8 | 1/8 | 1/4 | 1/4 | 1/2 | 1 | 2 | 3 | 5 | 8 | 14 |

# WEIGHT TABLE—FORGED STEEL FITTINGS

POUNDS EACH

| ASTMA – 105 | NORMAL PIPE SIZE – INCHES | | | | | | | | | | | |
|---|---|---|---|---|---|---|---|---|---|---|---|---|
| | 1/8 | 1/4 | 3/8" | 1/2 | 3/4 | 1 | 1-¼ | 1-½ | 2 | 2-½ | 3 | 4 |
| **Steel Screwed Bull Plugs:** | | | | | | | | | | | | |
| Hex Head Plug | 1/8 | 1/8 | 1/8 | 1/4 | 1/4 | 1/2 | 1 | 1 | 2 | 4 | 6 | 13 |
| Square Head Plug | 1/8 | 1/8 | 1/8 | 1/8 | 1/4 | 1/4 | 1/2 | 1 | 2 | 2 | 3 | 7 |
| Round Head Plug | 1/8 | 1/8 | 1/8 | 1/4 | 1/2 | 3/4 | 1 | 2 | 3 | 5 | 8 | 13 |
| Flush Bushing | – | 1/8 | 1/8 | 1/8 | 1/8 | 1/8 | 1/8 | 1/4 | 1/2 | 1/2 | 1 | 2 |

| Size Inches | Wt. |
|---|---|
| **Hex Head Bushings:** | |
| 1/4 x 1/8 | 1/8 |
| 3/8 x 1/4 & less | 1/8 |
| 1/2 x 3/8 & less | 1/8 |
| 3/4 x 1/2 & less | 1/8 |
| 1 x 3/4 & less | 1/8 |
| 1-1/4 x 1 & less | 1/2 |
| 1-1/2 x 1-1/4 & less | 1 |
| 2 x 3/4 & less | 2 |
| 2 x 1 & more | 2 |
| 2-1/2 x 1-1/4 & less | 2 |
| 2-1/2 x 1-1/2 & more | 2 |
| 3 x 1-1/2 & less | 4 |
| 3 x 2 & more | 4 |
| 4 x 1-1/2 & less | 8 |
| 4 x 2 & more | 8 |

# WEIGHT TABLE—FORGED STEEL FITTINGS

POUNDS EACH

| | NOMINAL PIPE SIZE | | | | | | | | | | | |
|---|---|---|---|---|---|---|---|---|---|---|---|---|
| ASTMA – 105 | 1/8" | 1/4" | 3/8" | 1/2" | 3/4" | 1" | 1-¼" | 1-½" | 2" | 2-½" | 3" | 4" |
| **2000-Pound Socket Welding Fittings** | | | | | | | | | | | | |
| 90° Elbows | 1/8 | 1/8 | 1/4 | 1/2 | 3/4 | 1 | 2 | 2 | 3 | 6 | 10 | 21 |
| 45° Elbows | 1/8 | 1/8 | 1/4 | 1/4 | 1/2 | 1 | 1 | 2 | 3 | 7 | 10 | 18 |
| Tees | 1/4 | 1/4 | 1/4 | 1/2 | 1 | 1 | 2 | 3 | 4 | 8 | 12 | 27 |
| Crosses | 1/4 | 1/4 | 1/2 | 3/4 | 1 | 2 | 2 | 3 | 5 | 18 | 23 | 40 |
| Laterals | – | 1/4 | 1/2 | 1 | 2 | 3 | 4 | 4 | 7 | – | – | – |
| Couplings | 1/8 | 1/8 | 1/8 | 1/4 | 1/4 | 1/2 | 3/4 | 1 | 2 | 3 | 4 | 7 |
| Reducing Couplings | – | 1/8 | 1/8 | 1/4 | 1/4 | 1/2 | 3/4 | 1 | 2 | 3 | 4 | 7 |
| Half Couplings | 1/8 | 1/8 | 1/8 | 1/8 | 1/4 | 1/4 | 1/2 | 1/2 | 3/4 | 1 | 2 | 4 |
| Caps | 1/8 | 1/8 | 1/8 | 1/4 | 1/4 | 1/2 | 3/4 | 1 | 2 | 3 | 4 | 7 |
| **3000-Pound Socket Welding Fittings** | | | | | | | | | | | | |
| 90° Elbows | – | 1/8 | 1/4 | 1/2 | 3/4 | 1 | 2 | 2 | 4 | 6 | 11 | 24 |
| 45° Elbows | – | 1/8 | 1/4 | 1/2 | 1/2 | 3/4 | 1 | 2 | 3 | 7 | 12 | 20 |
| Tees | – | 1/4 | 1/4 | 1/2 | 3/4 | 1 | 2 | 3 | 4 | 9 | 14 | 28 |
| Crosses | – | 1/4 | 1/4 | 1 | 1 | 2 | 2 | 3 | 6 | 16 | 20 | 32 |
| Laterals | – | 1/4 | 1/2 | 1 | 2 | 3 | 4 | 5 | 8 | – | – | – |
| Couplings | – | 1/8 | 1/8 | 1/4 | 1/2 | 3/4 | 1 | 1 | 2 | 3 | 5 | 9 |
| Reducing Couplings | – | 1/8 | 1/8 | 1/4 | 1/2 | 3/4 | 1 | 1 | 2 | 3 | 5 | 9 |
| Half Couplings | – | 1/8 | 1/8 | 1/8 | 1/4 | 1/2 | 1/2 | 1/2 | 1 | 1 | 2 | 4 |
| Caps | – | 1/8 | 1/4 | 1/4 | 1/2 | 1/2 | 1 | 1 | 2 | 3 | 5 | 8 |

# WEIGHT TABLE—FORGED STEEL FITTINGS

POUNDS EACH

4000 Pound Socket Welding Fittings:

| ASTM A-105 | NOMINAL PIPE SIZE | | | | | | | | | |
|---|---|---|---|---|---|---|---|---|---|---|
| | 3/8" | 1/2" | 3/4" | 1" | 1¼" | 1½" | 2" | 2½" | 3" | 4" |
| 90° Elbows | – | 1 | 1 | 2 | 3 | 5 | 7 | 12 | 19 | 26 |
| 45° Elbows | – | 1 | 1 | 2 | 2 | 4 | 5 | 10 | 14 | 27 |
| Tees | – | 1 | 2 | 3 | 4 | 6 | 8 | 16 | 24 | 33 |
| Crosses | – | 1 | 2 | 4 | 5 | 9 | 10 | 20 | 27 | 39 |
| Laterals | – | 2 | 3 | 5 | 6 | 12 | – | – | – | – |
| Couplings | – | 1/4 | 1/2 | 1 | 1 | 2 | 4 | 6 | 7 | 13 |
| Reducing Couplings | – | 1/4 | 1/2 | 1 | 1 | 2 | 4 | 6 | 7 | 13 |
| Half Coupling | – | 1/4 | 1/4 | 1/2 | 1/2 | 1 | 2 | 3 | 3 | 6 |
| Caps | – | 1/4 | 1/2 | 1 | 1 | 2 | 4 | 5 | 6 | 14 |

6000 Pound Socket Welding Fittings:

| ASTM A-105 | NOMINAL PIPE SIZE | | | | | | | | | |
|---|---|---|---|---|---|---|---|---|---|---|
| | 3/8" | 1/2" | 3/4" | 1" | 1¼" | 1½" | 2" | 2½" | 3" | 4" |
| 90° Elbows | 1/2 | 3/4 | 1 | 2 | 3 | 5 | 6 | 12 | 20 | 36 |
| 45° Elbows | 1/2 | 3/4 | 1 | 2 | 3 | 5 | 6 | 7 | 12 | 27 |
| Tees | 1 | 1 | 2 | 3 | 4 | 8 | 9 | 17 | 24 | 45 |
| Crosses | 1 | 1 | 2 | 4 | 5 | 9 | 10 | 20 | 30 | 42 |
| Laterals | 1 | 2 | 3 | 5 | 7 | 13 | – | – | – | – |
| Couplings | 1/4 | 1/2 | 3/4 | 2 | 2 | 3 | 5 | 6 | 9 | 16 |
| Reducing Couplings | 1/4 | 1/2 | 3/4 | 2 | 2 | 3 | 5 | 6 | 9 | 16 |
| Half Couplings | 1/4 | 1/4 | 1/2 | 1 | 1 | 2 | 3 | 3 | 5 | 8 |
| Caps | 1/4 | 1/4 | 1/2 | 1 | 1 | 2 | 3 | 6 | 8 | 13 |

# WEIGHT TABLE—MALLEABLE FITTINGS

NET WEIGHT PER HUNDRED PIECES

| | NOMINAL PIPE SIZE | | | | | | | | | | | |
|---|---|---|---|---|---|---|---|---|---|---|---|---|
| | 1/8" | 1/4" | 3/8" | 1/2" | 3/4" | 1" | 1-1/4" | 1-1/2" | 2" | 2-1/2" | 3" | 4" |
| **Standard Malleable Fittings:** | | | | | | | | | | | | |
| 90° Elbows | 7 | 11 | 17 | 25 | 42 | 60 | 110 | 140 | 230 | 380 | 530 | 960 |
| 90° Reducing Ells | – | 10 | 14 | 21 | 34 | 49 | 81 | 110 | 200 | 380 | 540 | 1100 |
| 45° Elbows | 7 | 11 | 17 | 23 | 36 | 52 | 91 | 120 | 210 | 320 | 450 | 800 |
| 90° Street Ells | 7 | 11 | 16 | 25 | 40 | 60 | 100 | 140 | 230 | 350 | 570 | 980 |
| Full Tees | 10 | 16 | 23 | 36 | 56 | 86 | 150 | 190 | 300 | 470 | 700 | 1200 |
| Reducing Tees | – | 14 | 21 | 31 | 51 | 76 | 120 | 150 | 210 | 380 | 550 | 1100 |
| Crosses | 12 | 19 | 28 | 43 | 68 | 97 | 170 | 210 | 350 | 530 | 770 | 1400 |
| Caps | – | 6 | 9 | 14 | 23 | 34 | 57 | 77 | 130 | 190 | 280 | 450 |
| Reducers | – | 8 | 11 | 16 | 26 | 44 | 65 | 87 | 140 | 220 | 390 | 680 |
| **300-Pound Malleable Fittings:** | | | | | | | | | | | | |
| 90° Elbows | – | 18 | 27 | 54 | 77 | 120 | 190 | 260 | 380 | 600 | 900 | 1700 |
| 45° Elbows | – | 21 | 34 | 49 | 70 | 110 | 190 | 230 | 350 | 580 | 890 | 1400 |
| 90° Street Ells | – | 13 | 23 | 45 | 68 | 110 | 180 | 240 | 370 | 640 | 1100 | – |
| Tees | – | 25 | 37 | 77 | 110 | 160 | 260 | 360 | 500 | 840 | 1300 | 2300 |
| Crosses | – | 36 | 52 | 93 | 140 | 200 | 300 | 410 | 620 | 1000 | 1500 | 2800 |

# SURFACE AREA OF INSULATED PIPE

SQUARE FOOT PER LINEAR FOOT

| Pipe Size Inches | Insulation Thickness | | | |
|---|---|---|---|---|
| | 1″ | 1-1/2″ | 2″ | 2-1/2″ |
| 1-1/2 | 0.96 | 1.30 | 1.55 | 1.82 |
| 2 | 1.12 | 1.40 | 1.87 | 1.95 |
| 3 | 1.47 | 1.70 | 1.98 | 2.25 |
| 4 | 1.80 | 2.00 | 2.24 | 2.50 |
| 6 | 2.35 | 2.55 | 2.82 | 3.07 |
| 8 | 2.95 | 3.08 | 3.34 | 3.60 |
| 10 | 3.55 | 3.64 | 3.90 | 4.16 |
| 12 | 4.16 | 4.30 | 4.42 | 4.68 |
| 14 | 4.55 | 4.69 | 4.83 | 5.10 |
| 16 | 4.97 | 5.11 | 5.24 | 5.50 |
| 18 | 5.40 | 5.54 | 5.72 | 5.93 |
| 20 | 6.05 | 6.19 | 6.30 | 6.55 |

# COATING COVERING CAPACITY

SQUARE FEET PER UNITS LISTED

| Material | Unit | Square Feet |
|---|---|---|
| Hot Asphalt | Pound | 3 |
| Asphalt Paint | Gallon | 100 |
| Asphalt Varnish | Gallon | 150 |
| Asphalt Emulsion | Gallon | 70 |
| Aluminum Paint | Gallon | 150 |
| Coldwater (5 pounds per gallon) | Gallon | 100 |
| Glue Sizing (2 pounds per gallon) | Gallon | 200 |
| Lap Cement | Gallon | 50 |
| Lead and Oil Paint | Gallon | 200 |
| Primer | Gallon | 150 |
| Asbestos | Gallon | 100 |

# SURFACE AREA OF PIPE FOR PAINTING

| Nominal Size Inches | Surface Area S.F. Per L.F. | Nominal Size Inches | Surface Area S.F. Per L.F. |
|---|---|---|---|
| 1 | 0.344 | 22 | 5.75 |
| 1-1/2 | 0.497 | 24 | 6.28 |
| 2 | 0.622 | 26 | 6.81 |
| 2-1/2 | 0.753 | 28 | 7.32 |
| 3 | 0.916 | 30 | 7.85 |
| 3-1/2 | 1.047 | 32 | 8.38 |
| 4 | 1.178 | 34 | 8.89 |
| 5 | 1.456 | 36 | 9.42 |
| 6 | 1.734 | 38 | 9.96 |
| 8 | 2.258 | 40 | 10.46 |
| 10 | 2.810 | 42 | 11.00 |
| 12 | 3.142 | 44 | 11.52 |
| 14 | 3.67 | 46 | 12.03 |
| 16 | 4.19 | 48 | 12.57 |
| 18 | 4.71 | 54 | 14.13 |
| 20 | 5.24 | 60 | 15.71 |